Environmental Toxicity and Human Health: Insights from Research

EDITORS

DR. POOJA KUMARI
ASSISTANT PROFESSOR
D. S. COLLEGE, ALIGARH

DR. B. R. SINGH
ASSISTANT PROFESSOR
D. S. COLLEGE, ALIGARH

notionpress.com

INDIA • SINGAPORE • MALAYSIA

ISBN
Paperback 979-8-90054-556-1
Hardcase 979-8-90054-557-8

Contents

Acknowledgement

We acknowledge the important research contributions of Dr. Barkha Shakya & Dr. Sonam Shakya on the toxic effects of arecoline, as well as Deeksha Singh & Dr. Sonal Singh for their integrated analysis of global environmental pollution. We thank Harikesh Kumar Singh, Zubaida Khatoon, Dr. Shalini Singh, Dr. Hadiya Husain & Alok Yadav for their contributions to studies on N-nitrosodiethylamine and liver disease progression, and Dr. Preeti Singh & Dr. Dharmendra Singh for their research on heavy metal contamination.

Special thanks to Irfaana Chowdhary & Prof. Mohammad Amir for their work on environmental pollutants including heavy metals, pesticides, and microplastics. We also appreciate the contribution of Dr. Neetu Nimesh, Varsha & Dr. Sangeeta Pal for highlighting the biodiversity of Keoladeo National Park. We acknowledge Prof Arif Ahmad & Dr. Nighat Gull for her research on hepatocellular carcinoma models, and Prachi Singh, Ms. Neha Rajpoot, Ms. Ragni Singh, & Dr. Fahad Ali for their review on monosodium glutamate and its health implications. Our sincere thanks go to Rajan Shukla & Dr. Ainul H. Ansari for their work linking geology and ecotoxicology.

We extend our gratitude to Rajat Kumar Gautam, Prof. Mukesh Kumar Bhardwaj, Sangram Singh, Dr. Avadhesh Kumar & Dr. Chandra Kant for their study on heavy metal contamination and remediation strategies. We also thank Dr. Renu Kumari & Dr. Manju Bhaskar for their contribution on sustainable pollution control techniques. We appreciate the innovative work of Sushobhit Varshney & Priyanshi Saraswat on converting cellulolytic waste into valuable fertilizer using Eisenia fetida. Our thanks are also due to Dr. Sonam Shakya, Dr. Barkha Shakya & Dr. Ishaat M. Khan for their research on ecotoxicological effects on ecosystems. We are grateful to

Dr. Surendra Pal Singh for his insights into biodiversity conservation within the Indian knowledge system.

The editors sincerely appreciate all contributors for their dedication and collaboration, which made this volume a valuable contribution to environmental science and sustainability. We also thank everyone who supported this work.

Our heartfelt thanks to Dr. Ashutosh Tiwari for his contribution on environmental conservation.

Regards,

Dr. Pooja Kumari & Dr. B. R. Singh

Chapter - 1

ROLE OF ACADEMIC INSTITUTES; AWARENESS AND CAPACITY BUILDING IN ENVIRONMENTAL CONSERVATION

Dr. Ashutosh Tiwari

Department of Education, Dharma Samaj College, Aligarh (202001), India

Corresponding email: ashutoshcari@gmail.com

ABSTRACT

Academic institutes play a pivotal and transformative role in shaping environmental consciousness and strengthening societal capacity for sustainable development in an era marked by accelerating ecological crises. Universities, colleges, research centres and professional institutes function as critical knowledge ecosystems where scientific understanding, ethical values and practical competencies related to environmental conservation are cultivated. Through structured teaching-learning processes, interdisciplinary research and innovation-driven practices, these institutions significantly influence how individuals and societies perceive, engage with and respond to environmental challenges.

This chapter critically examines the multifaceted contributions of academic institutions in promoting environmental awareness and building both individual and institutional capacities for conservation. Drawing upon global experiences and the Indian context, it explores the integration of environmental themes within curricula, the role of academic research in developing sustainable technologies and the importance of experiential learning approaches. The chapter also highlights the significance of community outreach programmes, teacher education, and extension activities in translating academic knowledge into grassroots environmental action. Furthermore, it discusses the growing engagement of academic institutions with policy processes, emphasizing their role in evidence-based decision-making and environmental governance. By analyzing theoretical frameworks and practical case studies, the chapter argues that academic institutions function as catalysts of environmental stewardship, acting as vital bridges between science, society and public policy. Through nurturing environmentally responsible values, critical thinking and leadership skills, academic institutions contribute decisively to the development of informed citizens and competent professionals capable of addressing present and future environmental challenges. The chapter concludes by emphasizing that strengthening the institutional capacity of academia is essential for achieving long-term ecological sustainability and intergenerational environmental justice.

Keywords: *Environmental Awareness, Capacity Building, Sustainable Development, Higher Education, Environmental Governance, Community Engagement*

1. INTRODUCTION

Environmental degradation has emerged as one of the most pressing global challenges of the twenty-first century, manifesting in climate change, biodiversity loss, desertification, water scarcity and pollution. These challenges require not only technological solutions but also a deep transformation in human values, knowledge systems and societal behaviour (Tilbury, 1995). In this context, academic institutes function as powerful agents of change by cultivating environmental awareness and building the capacities necessary for sustainable resource management.

Academic institutions are uniquely positioned because they operate at the intersection of knowledge creation, human resource development and public engagement. Through structured education, research, and outreach, these institutions shape how individuals perceive, understand, and respond to environmental challenges (Cortese, 2003). This chapter systematically analyzes how academic institutes contribute to environmental conservation through awareness building and capacity-enhancing interventions, with a focus on both global trends and the Indian higher education landscape.

2. CONCEPTUAL FOUNDATIONS

2.1. Environmental Awareness

Environmental awareness refers to the knowledge, attitudes, values and behaviours that support ecological sustainability and the responsible use of natural resources (UNESCO, 1977). It is not limited to cognitive understanding of environmental issues such as climate change, biodiversity loss and pollution; rather, it encompasses emotional, ethical and cultural dimensions that shape how individuals relate to nature. A person who is environmentally aware not only understands the scientific causes of environmental degradation but also develops a sense of personal responsibility, empathy toward living organisms, and a commitment to environmentally sound behaviour.

Academic institutes play a central role in nurturing environmental awareness by providing structured learning environments where scientific literacy and critical thinking are systematically developed. Through formal courses, interdisciplinary modules, seminars and project-based learning, students are exposed to contemporary environmental challenges and sustainable solutions (Tilbury, 1995). Experiential learning opportunities such as field visits, ecosystem mapping, nature camps, laboratory work and community-based projects further deepen environmental sensitivity by connecting theoretical knowledge with real-world ecological contexts.

Moreover, academic institutions foster environmental awareness by promoting dialogue, debate and reflective thinking on environmental ethics, intergenerational equity and the intrinsic value of nature. This helps learners move beyond anthropocentric worldviews toward more eccentric and sustainability-oriented perspectives. Research indicates that such holistic approaches to environmental education significantly influence long-term pro-environmental attitudes and behaviours among students (Kollmuss & Agyeman, 2002). Thus, environmental awareness within academic contexts becomes not merely an educational outcome but a foundational element of responsible citizenship and sustainable living.

2.2. Capacity Building

Capacity building is the systematic process of developing individual competencies, institutional frameworks, and systemic capabilities that enable effective environmental action and governance (UNDP, 2009). In the context of environmental conservation, capacity building extends beyond technical skill development to include strengthening organizational structures, leadership qualities, community engagement abilities and policy implementation mechanisms. It ensures that individuals and institutions are not only aware of environmental problems but are also equipped with the practical tools and strategic abilities required to address them.

Academic institutes contribute significantly to capacity building by offering specialized training programmes, professional courses, and research opportunities in environmental science, sustainable development, climate policy, renewable energy and natural resource management. These programmes develop technical expertise in areas such as environmental impact assessment, biodiversity monitoring, waste management and water conservation. Additionally, universities and research centres play a crucial role in strengthening institutional capacity by developing standard operating procedures, governance models and policy frameworks that support sustainable environmental management (OECD, 2012).

Capacity building also involves fostering leadership and change-making abilities among students, educators, and administrators. Through workshops, internships, field-based training and participatory projects, academic institutions cultivate problem-solving skills, collaborative competencies, and ethical leadership qualities. These efforts enhance the ability of individuals and organizations to adapt to emerging environmental challenges, implement innovative solutions, and promote resilient socio-ecological systems. Consequently, capacity building through academic institutions becomes a key driver of long-term environmental sustainability and effective conservation practice.

3. ACADEMIC INSTITUTES AS CENTRES OF ENVIRONMENTAL EDUCATION

3.1. Curriculum Integration

One of the primary and most impactful roles of academic institutions in environmental conservation is the systematic integration of environmental themes into formal curricula across diverse disciplines. Environmental education is no longer confined to specialized courses in environmental science; instead, concepts such as sustainable development, climate change, biodiversity conservation, ecosystem services and green technologies are increasingly embedded within subjects such as economics, sociology, engineering, management, law and the humanities (Sterling, 2001). This interdisciplinary curriculum design helps students understand that environmental challenges are not isolated scientific problems but are deeply interconnected with social, economic, political and cultural systems.

In the Indian context, a landmark step was taken when the University Grants Commission (UGC), following directives from the Supreme Court of India, made Environmental Studies a compulsory subject at the undergraduate level (UGC, 2003). This policy significantly broadened the reach of environmental awareness by ensuring that students from all streams-arts, science, commerce and professional courses-are exposed to fundamental environmental concepts. The inclusion of Environmental Studies has promoted ecological literacy, ethical sensitivity and a sense of environmental responsibility among millions of students across the country.

Furthermore, academic institutions increasingly emphasize outcome-based education, where learning objectives include the development of critical thinking, problem-solving and ethical reasoning in relation to environmental issues. Through case studies, project work, and interdisciplinary assignments, students are encouraged to analyze real-world environmental problems such as urban pollution, water scarcity, deforestation and climate vulnerability. This holistic curricular approach fosters a deeper understanding of the complex relationships between human societies and natural ecosystems, thereby laying a strong intellectual foundation for sustainable behaviour and professional practice.

3.2. Experiential and Field-Based Learning

In addition to classroom-based instruction, academic institutions play a crucial role in promoting environmental awareness through experiential and field-based learning methodologies. These approaches emphasize learning by doing and direct engagement with natural and socio-ecological systems. Field visits to forests, wetlands, wildlife sanctuaries, coastal ecosystems and biodiversity hotspots provide students with firsthand exposure to ecological processes and conservation challenges. Such experiences help bridge the gap between theoretical knowledge and real-world environmental realities.

Academic institutions also facilitate ecological surveys, biodiversity assessments, water quality testing, and citizen science projects, which actively involve students in data collection, observation and analysis. Participation in these activities enhances scientific skills and fosters a sense of ownership and responsibility toward the environment. Internship programmes with environmental non-governmental organizations (NGOs), forest departments, pollution control boards and international agencies further strengthen students' practical competencies and professional orientation towards environmental careers.

Research evidence strongly supports the effectiveness of experiential learning in environmental education. Studies indicate that direct interaction with natural environments significantly enhances environmental empathy, attitudes and long-term pro-environmental behaviour compared to conventional classroom instruction alone (Rickinson, 2001). Experiential learning also promotes teamwork, leadership and reflective thinking, which are essential attributes for effective environmental stewardship. By embedding field-based and experiential learning within academic programmes, institutions ensure that environmental education becomes transformative rather than merely informative, ultimately contributing to the development of environmentally responsible and socially conscious citizens.

4. ROLE IN RESEARCH AND KNOWLEDGE GENERATION

4.1. Environmental Research and Innovation

Universities and research institutes are hubs of scientific inquiry and innovation. They conduct foundational and applied research in areas such as climate change mitigation, renewable energy, pollution control, waste management and biodiversity conservation (Clark, 2007).

Academic research contributes by:

- Identifying environmental problems and their root causes
- Developing sustainable technologies
- Providing evidence-based policy recommendations

This research-driven knowledge strengthens societal capacity to respond to environmental challenges.

4.2. Interdisciplinary Research Culture

Environmental problems are inherently complex and require interdisciplinary solutions. Academic institutes increasingly adopt interdisciplinary research models that integrate natural sciences, social sciences, economics, and humanities (Pretty, 2008). Such collaboration enhances the capacity of societies to develop adaptive and resilient environmental governance frameworks.

5. BUILDING HUMAN CAPITAL FOR ENVIRONMENTAL CONSERVATION

5.1. Professional Training and Skill Development

Academic institutions play a crucial and strategic role in building human capital for environmental conservation by preparing a wide range of environmental professionals. These include environmental scientists and engineers, wildlife biologists and foresters, climate policy analysts, sustainability managers, urban planners and renewable energy specialists. Through structured degree programmes at undergraduate, postgraduate and doctoral levels, as well as diplomas and short-term certification courses, academic institutions systematically develop the technical, analytical and ethical competencies required for effective conservation practice (Leal Filho, 2011).

These programmes equip students with specialized knowledge in areas such as environmental impact assessment, biodiversity monitoring, geographic information systems (GIS), remote sensing, pollution control technologies, waste management, water resource management, climate risk assessment and ecological restoration. Alongside technical training, institutions increasingly emphasize the development of managerial and leadership skills, including project planning, stakeholder coordination, financial management and policy analysis. This dual focus ensures that graduates are not only scientifically competent but also capable of designing, implementing, and managing complex conservation projects.

Furthermore, academic institutions foster innovation and problem-solving through laboratory-based research, fieldwork, internships and industry collaborations. Exposure to real-world environmental challenges strengthens professional competence and enhances employability in governmental agencies, non-governmental organizations, international development bodies and private sustainability sectors. In this way, academic training transforms students into skilled practitioners who can contribute effectively to conservation, climate adaptation, and sustainable development initiatives at local, national and global levels.

5.2. Teacher Education and Multiplying Impact

Teacher education institutions and colleges of education occupy a unique and highly influential position in amplifying environmental awareness and conservation values across society. Teachers act as powerful "multipliers" because a single well-trained teacher can influence the environmental knowledge, attitudes and behaviours of hundreds of students over the course of their professional life (Palmer, 1998). When environmental education is systematically incorporated into teacher preparation programmes, it creates a long-lasting and expansive ripple effect that extends far beyond the classroom.

Teacher training in environmental education focuses not only on subject content but also on appropriate pedagogical strategies, such as inquiry-based learning, project-based learning, outdoor education, and community engagement. Future teachers are trained to integrate environmental themes across subjects, linking science, social studies, language and ethics in meaningful ways. This interdisciplinary teaching approach encourages students to view environmental issues as integral to their daily lives rather than as abstract or isolated topics.

Additionally, teacher education institutions promote reflective practice and value-based education, enabling future educators to model environmentally responsible behaviour in their personal and professional lives. Teachers trained in sustainability principles are more likely to organize school-level activities such as tree plantation drives, waste reduction campaigns, water conservation projects and eco-clubs. Through this sustained influence, teacher education institutions help institutionalize environmental values within the school system, ensuring that sustainability consciousness is transmitted across generations and embedded in the broader culture of society.

6. ACADEMIC INSTITUTES AND COMMUNITY ENGAGEMENT

6.1. Extension and Outreach Programmes

Many academic institutions undertake community outreach activities that extend environmental awareness beyond campus boundaries. These include:

- Village-level awareness campaigns
- Workshops on waste management and water conservation
- Tree plantation drives and biodiversity conservation campaigns

Such programmes transform universities into "living laboratories" for sustainable practices (Brundiers & Wiek, 2011).

6.2. Participatory Action Research

Participatory action research involves collaboration between academic researchers and local communities to address real-world environmental problems (Reason & Bradbury, 2008). This approach enhances community capacity by combining scientific knowledge with local wisdom, thereby ensuring culturally appropriate and socially acceptable conservation strategies.

7. ROLE IN POLICY ADVOCACY AND ENVIRONMENTAL GOVERNANCE

7.1. Evidence-Based Policy Support

Academic institutions support governments and international agencies by providing research-based policy inputs. Expert committees, advisory panels and white papers prepared by academic experts have influenced environmental legislation and governance structures worldwide (Jasanoff, 2004).

In India, academic institutes have contributed to policy frameworks such as:

- National Environmental Policy (2006)
- National Action Plan on Climate Change (NAPCC)
- Biodiversity Action Plans

7.2. Training of Administrators and Decision-Makers

Capacity building is not limited to students. Academic institutions conduct training programmes for:

- Environmental administrators
- Forest officers

- Local governance bodies
- Industrial managers

These programmes improve governance quality and strengthen the enforcement of environmental regulations (OECD, 2012).

8. ACADEMIC CAMPUSES AS MODELS OF SUSTAINABILITY

8.1. Green Campus Initiatives

Universities are increasingly transforming their campuses into models of sustainable living by adopting:

- Solar energy systems
- Rainwater harvesting structures
- Solid waste management and recycling
- Green buildings and biodiversity parks

By practicing what they teach, academic institutions reinforce environmental values through lived experience (Shriberg, 2002).

8.2. Institutional Culture and Environmental Ethics

Sustainable campus initiatives cultivate an institutional culture of environmental responsibility. Students and staff develop environmentally friendly habits that often continue beyond their academic lives, contributing to broader societal change (Lozano et al., 2013).

9. CHALLENGES FACED BY ACADEMIC INSTITUTES

Despite their potential, academic institutions face multiple constraints in fully realizing their role in environmental conservation:

- Inadequate funding for environmental research
- Fragmented curricula and lack of interdisciplinary integration
- Limited faculty training in sustainability education

- Weak linkage between academic research and policy implementation
- Institutional resistance to systemic transformation

These challenges require structural reforms, policy support and international collaboration to strengthen the role of academic institutes.

10. FUTURE DIRECTIONS AND RECOMMENDATIONS

To enhance the effectiveness of academic institutes in environmental awareness and capacity building, the following strategies are suggested:

1. **Mainstreaming Sustainability Across Disciplines**-Environmental perspectives should be integrated into arts, humanities, commerce and technology programmes.
2. **Strengthening Research-Policy Linkages**-Mechanisms for translating research into policy action must be institutionalized.
3. **Community-Embedded Learning**-Service-learning and problem-based learning models should be expanded.
4. **International Collaboration**-Global research networks can strengthen knowledge exchange.
5. **Institutional Reform**-Universities should adopt sustainability as a core institutional mission rather than an optional activity.

11. CONCLUSION

Academic institutes occupy a strategic and indispensable position in the global effort to conserve the environment and promote sustainable development. As centres of knowledge creation, critical inquiry, and human resource development, universities and other higher education institutions serve as powerful agents of environmental transformation. By generating scientific knowledge, cultivating environmental awareness, and building both individual and institutional capacities, academic institutions contribute significantly to shaping environmentally responsible societies.

The integration of environmental education into formal curricula across disciplines has enabled students from diverse academic backgrounds to develop ecological literacy and ethical sensitivity toward nature. Research-driven innovation conducted within academic institutions has contributed to the development of sustainable technologies, evidence-based policy frameworks and adaptive strategies for climate change and biodiversity conservation. Furthermore, community outreach and extension activities have strengthened the link between academic knowledge and grassroots environmental action, ensuring that conservation efforts are socially inclusive and locally relevant.

Academic institutes have also emerged as key contributors to environmental governance by supporting policy formulation, institutional strengthening and capacity building for government officials, community leaders and civil society organizations. Through these multifaceted roles, they demonstrate that they are not passive observers of environmental change, but active architects of environmental stewardship and sustainable futures. Their ability to bridge science, society and policy uniquely positions them to influence both present and future generations.

Strengthening the role of academic institutions is therefore indispensable for achieving long-term ecological sustainability, climate resilience and intergenerational environmental justice. This requires sustained investment in environmental research, interdisciplinary education, institutional reforms, and stronger collaboration between academia, government, industry and communities. Only through such integrated and forward-looking approaches can academic institutions continue to contribute meaningfully to the preservation of natural ecosystems and the creation of a more just, resilient and sustainable world.

REFERENCES

1. Brundiers, K. & Wiek, A. (2011). Educating students in real-world sustainability research: Vision and implementation. *Innovative Higher Education, 36*(2), 107–124.
2. Clark, W. C. (2007). Sustainability science: A room of its own. *Proceedings of the National Academy of Sciences, 104*(6), 1737–1738.
3. Cortese, A. D. (2003). The critical role of higher education in creating a sustainable future. *Planning for Higher Education, 31*(3), 15–22.
4. Jasanoff, S. (2004). *States of knowledge: The co-production of science and social order*. Routledge.
5. Leal Filho, W. (2011). About the role of universities and their contribution to sustainable development. *Higher Education Policy, 24*(4), 427–438.
6. Lozano, R. et al. (2013). Declarations for sustainability in higher education: Becoming better leaders, through addressing the university system. *Journal of Cleaner Production, 48*, 10–19.
7. OECD (2012). *Environmental governance in OECD countries: A review of strategic environmental assessment*. OECD Publishing.
8. Palmer, J. (1998). *Environmental education in the 21st century: Theory, practice, progress and promise*. Routledge.
9. Pretty, J. (2008). Agricultural sustainability: Concepts, principles and evidence. *Philosophical Transactions of the Royal Society B*, 363(1491), 447–465.
10. Reason, P. & Bradbury, H. (2008). *The SAGE handbook of action research: Participative inquiry and practice*. Sage Publications.
11. Rickinson, M. (2001). Learners and learning in environmental education: A critical review. *Environmental Education Research, 7*(3), 207–320.
12. Shriberg, M. (2002). Institutional assessment tools for sustainability in higher education. *International Journal of Sustainability in Higher Education, 3*(3), 254–270.

13. Sterling, S. (2001). *Sustainable education: Re-visioning learning and change*. Green Books.

14. Tilbury, D. (1995). Environmental education for sustainability: Defining the new focus of environmental education in the 1990s. *Environmental Education Research, 1*(2), 195–212.

15. Turner, B. L. et al. (2003). A framework for vulnerability analysis in sustainability science. *Proceedings of the National Academy of Sciences, 100*(14), 8074–8079.

16. UNDP (2009). *Capacity development: A UNDP primer*. United Nations Development Programme.

17. UNESCO (1977). *Tbilisi Declaration: Intergovernmental Conference on Environmental Education*. UNESCO.

18. UGC (2003). *Environmental education: Report of the UGC core group*. University Grants Commission, India.

Chapter - 2

SUSTAINABLE ENVIRONMENTAL MANAGEMENT: GREEN CHEMISTRY

Dr. Avadhesh Kumar[1*], Rajat Gautam[2] and Dr. Chandrakant[3]

[1]Assistant Professor, Department of Chemistry, Raja Mahendra Pratap Singh University, Aligarh-202001, U.P., India

[2]Scholar, Department of Botany, D. S. College, Aligarh-202001, U.P., India

[3]Assistant Professor, Department of Botany, D. S. College, Aligarh-202001, U.P., India

***Corresponding email: akkorganic2019@gmail.com**

ABSTRACT

Green chemistry has emerged as a leading interdisciplinary field integrating chemistry and nanotechnology to promote sustainable development. Its scope has expanded significantly in recent years, offering innovative and eco-friendly solutions across modern science and engineering. By overcoming the limitations of traditional synthetic methods, green chemistry introduces environmentally benign approaches for synthesizing nanoparticles and other materials, reducing the harmful impacts of hazardous chemicals. At its core, green chemistry is guided by well-defined principles aimed at minimizing the use and generation of toxic substances during chemical processes. Three fundamental principles underpin this approach: the use of greener solvents, the elimination of harmful by-products, and the maintenance of atom economy. These principles support the development of processes that are rapid, clean, safe, cost-effective, non-toxic and commercially viable, while ensuring high productivity and material purity. Green chemistry plays a crucial role in sustainable environmental development by addressing growing global concerns about pollution and resource depletion. Its applications extend to diverse fields, including biomedicine, agriculture, pharmaceuticals, biosensors, catalysis and wastewater treatment. The demand for biocompatible and sustainable materials in these areas has further accelerated the adoption of green synthetic strategies. In addition to industrial applications, green chemistry is increasingly emphasized in academic settings through laboratory experiments and skill development programs. Overall, this field provides a comprehensive framework for environmentally responsible innovation, contributing significantly to sustainable environmental management and agricultural advancement.

Keywords: *Green Chemistry, Greener Solvents, Atom Economy, Sustain Environment, Non-hazardous*

Graphical Abstract

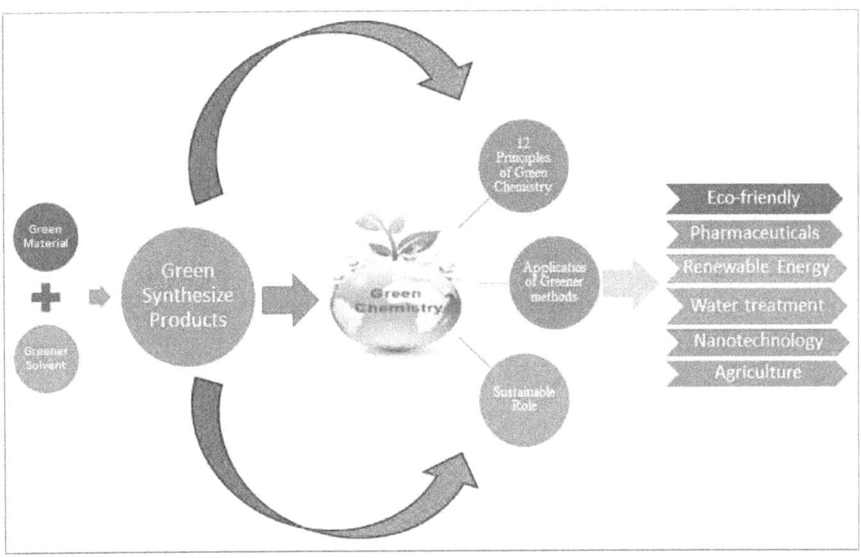

Features Highlight's

- This chapter study sustainable environment management and development through green chemistry.
- Signify the advancements search to processes that use greener chemicals and produce by-products while using minimal energy.
- Highlights the import of its use in daily life and show the well-being the utilization of green chemistry for society as well as ecosystem.
- This unique article addressing the wide interest in green and clean procedure approach, which has grown significantly in recent years.

1. INTRODUCTION

In the 19th century, environmentally conscious concepts such as clean chemistry, environmental chemistry, green chemistry and sustainable chemistry emerged in the field of chemistry. However, these terms were often debated and lacked clear definitions (Centi & Perathoner, 2003; Eissen & Metzger, 2002). Among them, green chemistry gained prominence as an approach focused on minimizing hazardous substances and environmental impact in chemical processes and product design. Paul Anastas and John Warner introduced the principles of green chemistry (GCP) to prevent pollution at its source. Their approach emphasized the use of safer chemicals, catalysts instead of stoichiometric reagents and waste reduction (Zhang, 2017; Anastas & Warner, 1998). By improving chemical manufacturing efficiency while enhancing environmental protection, green chemistry became a valuable framework for researchers, chemists and industrialists.

Green Chemistry Science

Greener chemistry is a scientific approach that combines biological catalysis with technological innovation to create environmentally and economically sustainable solutions, serving as a unique tool for building a resilient civilization. It seeks to transform the way chemicals are produced and used, ensuring safety and sustainability (Anastas, 2007). This vision aligns closely with broader sustainability goals and the terms "green chemistry" and "sustainable chemistry" are often used interchangeably. Achieving a sustainable civilization requires transformation in material and energy use, particularly through transmaterialization and dematerialization. Transmaterialization involves shifting from hazardous and exhaustible resources to safer and renewable alternatives. Dematerialization focuses on reducing material and energy inputs while maintaining economic growth and social well-being. Together, these strategies help minimize the environmental and health risks associated with chemical production. Future chemical products and processes must therefore be designed for safe manufacture, use, and disposal. Green chemistry supports this transition by embedding sustainability principles into research and industrial practice.

Green chemistry provides a rational scientific and technical framework for transforming industrial processes and reducing or eliminating by-products in chemical synthesis. By applying sustainability metrics, it identifies opportunities for innovation while minimizing unintended hazards. Its emergence was influenced by growing global environmental concerns, including the 1992 Rio Declaration and Agenda 21, which emphasized safer chemicals and shorter product life cycles.

Related concepts include atom economy, process optimization, inherently safer chemistry, life-cycle assessment, ionic liquids, alternative feedstocks, and renewable energy. The practical application of green chemistry is guided by the twelve principles established by Anastas and Warner (1998), forming the foundation for sustainable chemical development (Lozano et al., 2018; Wardencki et al., 2005).

Crucial or Vital point

A suitable way to quickly show (e.g. to students) the atrocity of the waste problem. Green chemistry is not only beneficial for the environment it is also benefit for business and human health.

2. WHAT IS "GREEN CHEMISTRY"

The term 'Green chemistry' can be defined as invention, design, development and sustained application of chemical products and process to reduce or eliminate the sources of pollutants or waste biproducts. To prevent generated pollutants by employs its green approach. Although the terms green also refers to practices that eliminate use of hazardous substances and materials, energy, water and other resorts as well as protect natural resorts through efficient use (EPA, 1996.) to human and environment system. Generally, in this advanced branch of science younger chemists are employ some principles to healthy, safe and clean environment.

Green chemistry represents the pillars that hold up our sustainable future. It is imperative to teach the value of green chemistry to tomorrow's chemist. By Daryle Busch (Ex-President, ACS).

3. EDUCATION OF GREEN CHEMISTRY

Several thoughts exist relative to the word environment sustainable development. Fundamentally, sustainable development is describing in a broadly statement: "the development that meets the needs of the present without comprising the abilities of future generations to meet their own needs" (World Commission on Environment and Development, 1987). Today teaching is primary for sustainability, as it effective the implementation to eco-friendly ideas, influence decision to build sustain society, and control the standard of life. Green teaching is promoting to conscious ecofriendly development that familiar as *education for sustainable development*. These are valuable about to sustainable development, that centred on human beings and asks humankind to change its daily lifestyle. The ultimate focus of education for sustainable development is to create a sense of responsibility that takes into account a guiding principle in the classroom learning as well as completing education, socially, economically, and environmentally outcome on human daily life (Burmeister et al., 2012). Consequently, recent teaching for sustainable development focused on active participation as a significance effect that a sense of responsibility to actively contribute to evolution a sustainable environmental way of living.

In current period, green chemistry is not a separate branch of science. Alternatively, it is a novel way of thinking about science in an accountable manner thereby the lives of future generations are no agreement from today's actions. Implementing of green chemistry content into syllabus provides a foundation for developing a sustainable chemical sector and fostering a sustainable society (Braun et al., 2006).

4. EVOLUTION AND IMPLEMENTATION OF SUSTAINABLE DEVELOPMENT EDUCATION

This section denotes where and how to implement the basics of green chemistry principles, methods in an education system like teacher education, secondary school curriculum and experimental laboratory to ecofriendly development, the contents of the syllabus and pedagogy applied to convey the subject matter are centralizing to the certain values, worldwide, and perspective that are stimulate (Fien, 2000). The authors recommend the curriculum to be holistic and pedagogy to be student centred. Both goals can be achieved when everyday life examples and knowledge are welcomed as the grounding for students' subsequent learning (Anderson, 2007). Integration of everyday life examples and knowledge leads to an interdisciplinary perspective on subject matter (Roth, 2003). This is so because sustainable system of knowledge connection gap between the information and application (David et al., 2003). Compatibly, ecofriendly development from green chemistry concepts also recognize teaching and learning to combine curriculum subjects, accordingly, keep up leaners to actively involves in interdisciplinary task statements and finding the answer (Yencken et al., 2000).

5. PRINCIPLES OF GREEN CHEMISTRY

The green chemistry objectives to remove or terminate the hazardous and toxic waste materials formed during the chemical producer and also be to prevent human health and environmental system by inventing a novel greener approach (Anastas et al., 1996). The following 12 principles give an effective and efficient route for young researchers to applying green chemistry.

"The ideal chemical process is that which a one-armed operator can perform by pouring the reactants into a bath tub and collecting pure product from the drain hole" Sir John Cornforth (Nobel Prize 1975).

The application of the green approach in practice is guided by 12 principles that establish a greener method of chemistry, as shown in Figure 1. These principles of green chemistry were initially proposed by Professor Paul Anastas and John Warner in 1998 as a novel approach, and have since been expanded to play a significant role in environmentally friendly development (Wardencki et al., 2005). They emphasize preventing pollutants at their source rather than relying on end-of-pipe treatment solutions. The theory and practice of these principles provide a roadmap that includes: preventing waste generation; improving atom economy; using safer and less hazardous substances in synthesis; designing safer chemicals; enhancing energy efficiency; using renewable feedstocks; reducing the need for derivatives; promoting catalysis; designing chemicals for degradation; enabling real-time analysis for pollution prevention; and ensuring inherently safer chemistry to reduce the risk of accidents (Anastas & Warner, 1998).

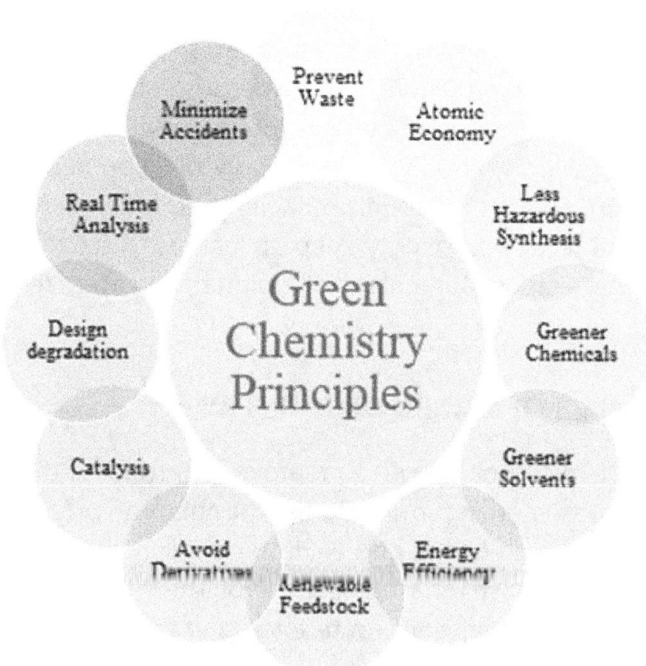

Figure 1: Principles of Green Chemistry

5.1. Prevent Waste

Imagine manufacturing industry that operate in a closed-loop system, where waste from one process becomes the feedstock for another. Green chemistry research is actively developing these cutting-edge techniques, reducing reliance on landfills and minimizing environmental strain. The reduction of costly raw materials should be ignored, and their incorporation into the last product of the reaction should be increase. The mass equilibrium of reactions and procedure steps must be estimate for optimization. This type of approach will make less waste products, alleviate environmentally harmful, and improve sustainability.

5.2. Atom Economy

By maximizing the use of starting materials, chemists can reduce waste and enhance reaction efficiency. This express to lower production costs, making green chemistry not only eco-friendly but also economically attractive and effective.

$$\text{Percent atom economy} = \frac{Molar\ weight\ of\ desired\ producta}{Molar\ weight\ of\ all\ reactant} \times 100$$

5.3. Less Hazardous Synthesis

Implementing safer synthetic methods not only enhances workplace safety for chemists but also significantly minimizes risks associated with transportation and storage. Additionally, it minimizes the potential for environmental contamination if spills or accidents occur.

5.4. Greener Chemicals

Selection of the greener chemicals directed to products designed with inherent safety in mind can revolutionize various industries. Imagine non-toxic cleaning products that are just as effective, or fire retardants that don't pose health risks. Green chemistry paves the way for a safer future.

5.5. Greener Solvents

The development of bio-originated solvents or even water-based alternatives can significantly reduce the environmental impact of chemical operation. These advancements not only minimize health risks but also open entries for more sustainable manufacturing aspects.

5.6. Energy Efficiency

Green chemistry shoulder of chemical reaction engineering, enabling the development of procedures that handle at room temperature and pressure conditions. This not only reduces energy consumption and greenhouse gas emissions but also simplifies reaction setups, potentially leading to more portable and decentralized chemical production.

5.7. Renewable Stocks

Conversion from fossil to renewable resources like bio-based materials or adopt carbon dioxide represents a significant step towards a sustainable future. Green chemistry is at the forefront of this transition, developing methods to utilize these renewable resources for chemical production.

5.8. Avoid Derivatives

Minimizing unnecessary steps in a synthesis not only reduces waste but also streamlines the process, potentially leading to faster production times and lower costs. Green chemistry is constantly seeking elegant solutions that achieve the desired outcome with minimal manipulation.

5.9. Catalysis

Catalysts are the workhorses of green chemistry. By using them strategically, chemists can achieve high reaction efficiency with minimal waste. Research in this area is ongoing, with scientists developing ever more powerful and selective catalysts for specific reactions. In this perspectives several type of catalyst such as bio-catalyst, heterogeneous, homogenous and organo catalyst are used to design and develop greener techniques shown in figure 2 *(Sheldon, Arends and Hanefeld, Green Chemistry & Catalysis, Wiley, New York, 2007).*

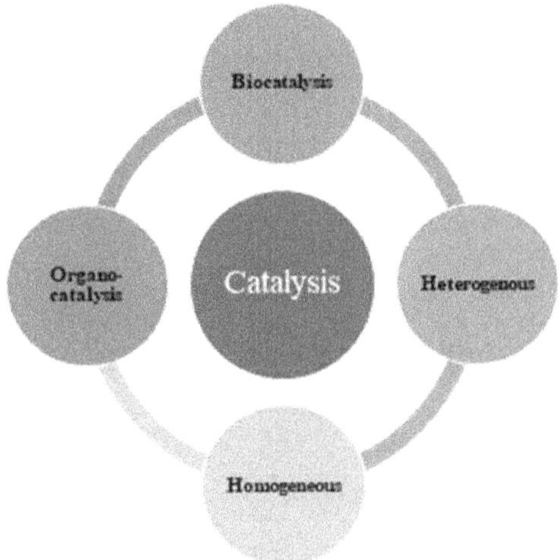

Figure 2: Important Principles of Green Chemistry

5.10. Designs for Degradation

Imagine a world where disposable products decompose into harmless components after use. This principle paves the way for the development of bioplastics and other biodegradable materials, significantly reducing plastic pollution and its negative impact on ecosystems.

5.11. Real-Time Analysis

Advanced monitoring systems can detect potential problems in real-time, allowing for immediate adjustments to prevent pollution formation. This not only safeguards the environment but also ensures consistent product quality and minimizes production downtime.

5.12. Accident Prevention

Design the safer substances and their state (solid, liquid, or gas) to minimize and minimize the risk of coincidences throughout the chemical lifecycle as well as explosions, fires and liberate in the environment. These shift to safer workplaces, lower insurance costs, and ultimately, a more sustainable chemical industry.

Spirit of Green Chemistry

P – Prevent wastes

R – Renewable materials

O – Omit derivatisation steps

D – Degradable chemical products

U – Use of safe synthetic methods

C – Catalytic reagents

T – Temperature, Pressure ambient

I – In-Process monitoring

V – Very few auxiliary substrates

E – E-factor, maximise feed in product

L – Low toxicity of chemical products

Y – Yes, it is safe

*(Source: S. L. Y. Tang, R. L. Smith and M. Poliakoff, Green Chem., **2005**, 7,761)*

6. METRICS OF GREEN CHEMISTRY

Some basic formulas are used to calculate factors in green chemistry

E-Factor	**Atom Efficiency**
E = $\frac{Toatl\ mass\ of\ waste}{Mass\ of\ final\ product}$	**AE (%)** = $\frac{Molar\ weight\ of\ product}{\Sigma Molar\ weight\ of\ reactants} \times 100$
Mass intensity (MI)	**Reaction mass efficiency (RME)**
MI = $\frac{Total\ mass\ of\ waste}{Mass\ of\ product}$	RME (%) = $\frac{Mass\ of\ product}{Mass\ of\ A + Mass\ of\ B} \times 100$
Mass productivity (MP)	**Carbon efficiency (CE)**
MP = $\frac{Mass\ of\ product}{Total\ mass\ in\ process}$	CE (%) = $\frac{Carbon\ in\ product}{Total\ carbon\ in\ reactants} \times 100$

Effective mass yield (EMY)	**The Environmental Impact EQ**
$\text{EMY (\%)} = \frac{\textit{Mass of product}}{\textit{Mass of hazardous reagents}}$	EQ = E (kg waste) × Q Q = Unfriendliness Multiplier e.g. NaCl: Q = 1 (arbitrary) Cr salts: Q = 1000? *(R.A.Sheldon, Chem. and Ind, 1992, 903 ; 1997, 12)*

E-Factor are calculated by E Factor = kg waste/kg product of some chemical substance

Products	Tonnage	E Factor
Oil Refining	106-108	<0.1
Bulk Chemicals	104-106	<1 - 5
Fine chemical Industry	102-104	5 - >50
Pharmaceutical Industry	10-103	25 - >100

"Another aspect of process development mentioned by all pharmaceutical process, chemists who spoke with CandEN is the need for determining an E Factor".

(Source: A. N. Thayer, CandEN, August 6, 2007, pp. 11-19, R.A.Sheldon, Chem and Ind, 1992, 903 ; 1997, 12)

7. TOXICITY OF SOLVENTS IN GREEN CHEMISTRY

Benzene	:	Highly carcinogenic
Carbon tetra chloride (CCL^4)	:	Highly carcinogenic
Chloroform (CHCl3)	:	Cancer suspect agent
Acetonitrile	:	Very toxic
Dioxane	:	Cancer suspect agent
Dimethyl furan (DMF)	:	Teratogen irritant

Tetrahydrofuran (THF)	:	Irritant
Ethyl acetate	:	Irritant

*(**Source:** Merck Index, Aldrich Catalogue Chemical Supplier agencies)*

7.1. Disadvantage of Chemical Solvents

When we using "greener" solvents provide environmental benefits, they can also have few disadvantages seen in figure 3, like high volatile, fire hazard, costly, toxicity, limited compatibility with certain samples, potentially decreased resolution and sensitivity in chromatography, and energy consumption in some applications.

Figure 3: Limitations of green chemistry solvents

7.2. Selection of Greener Solvent Guide

In the green synthesis solvent contribute important role play of chemical reaction and process. We can eliminate of hazardous solvent is the greenest options. The reactions process requires a optimize solvent, a number of substitute greener solvents have been employed. Although the replace of solvents in the reactions may be aided using another like, heat, microwave Irradiation has effectively encouraged a numerous of reactions without the adding solvent. The crucial classes of substituted solvents including ionic liquids, supercritical CO2/gas-expanded liquids, and switchable solvents. Green solvents are designed to minimize or eliminate exposures to volatile organic compounds, which contribute to air pollution, flammability hazards, and adverse human health effects.

To promote the adoption of alternative some green solvents in chemical industries, tools have been adopted to aid in the solvent choosing process, see in figure 4. These tools help chemists quickly identify safer alternatives for many of the common hazardous solvents.

Figure 4: Solvent replacement guide

(Source: P. Dunn et al, Green Chem. 2008, 10, 31-36)

Abbreviations: NMP, N-Methyl-2-pyrrolidone; DMF, Dimethylfuran, DMAc, Dimethylacetamide; MeTHF, Methyltetrahydrofuran; THF, Tetrahydrofuran; DMSO, Dimethylsulfoxide; MEK, Methylethylketone.

8. APPLICATION OF GREEN CHEMISTRY

Young researchers are increasingly developing renewable, biodegradable, and non-toxic materials to support sustainable innovation. Green chemistry plays a significant role across multiple sectors, including the pharmaceutical industry, energy science, eco-friendly dry cleaning, research laboratories, water purification, catalyst design and nanoscience applications. It contributes to achieving the United Nations Sustainable

Development Goals (SDGs) by promoting improved water stewardship, enhanced energy efficiency and environmentally responsible production systems. Through innovative chemical approaches, green chemistry enables non-toxic material flows and supports sustainable supply chains throughout a product's entire life cycle. Beyond environmental benefits, it also strengthens social responsibility by encouraging ethical practices, integrating sustainability into education and curricula and promoting social justice. Overall, green chemistry fosters environmental protection, economic viability and social well-being through safer and more sustainable chemical innovations.

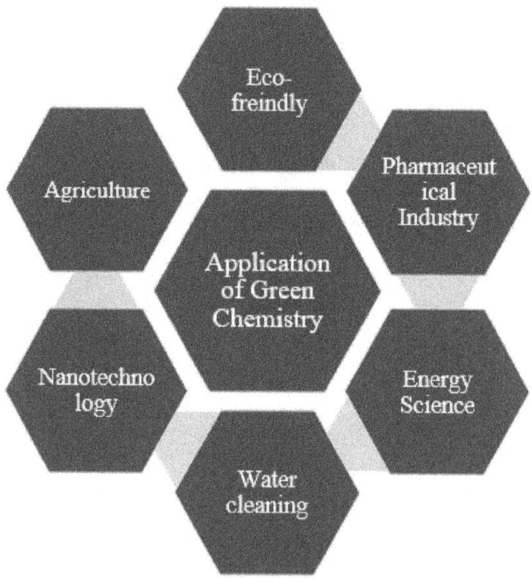

Figure 5: Application of green chemistry in various fields for sustained development

The development of a sustainable environment through green methods is illustrated with several significant examples. These preparations demonstrate how such approaches can be safe for human health while also being eco-friendly.

- *Preparation of 2-cyno, 3-phenyl, acrylic acid ethyl ester*

Conventional method:

In this method by using non-greener solvent toluene and piperidine is toxic and is not eco-friendly. Where potassium Fluorosilicate (KSF) is solid acid catalyst that is as renewable.

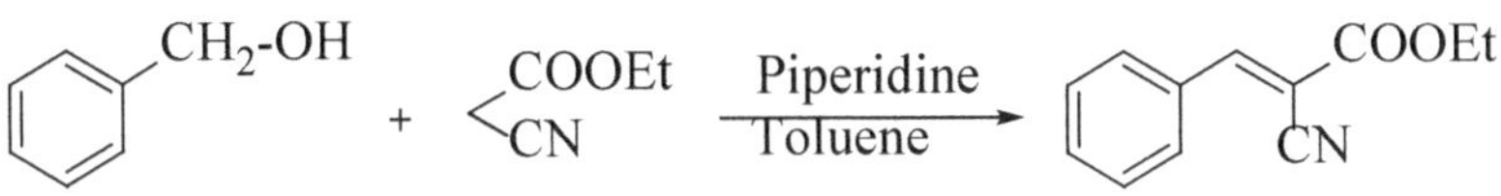

Greener approach:

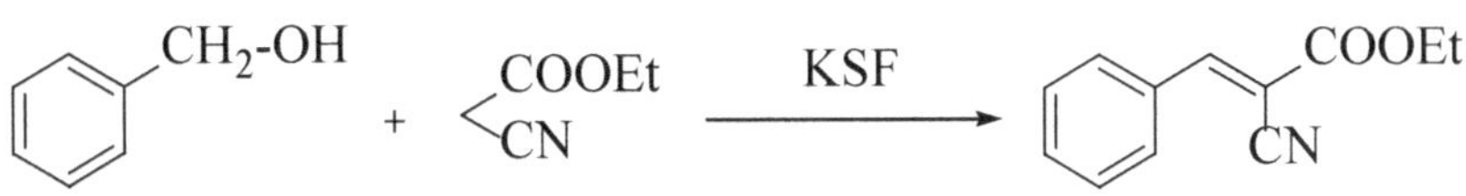

- *Synthesis of Phenol by nitration*

Conventional method:

When benzene react with non-green component in the presence of sulphuric acid in conventional method than gives p-nitro phenol,

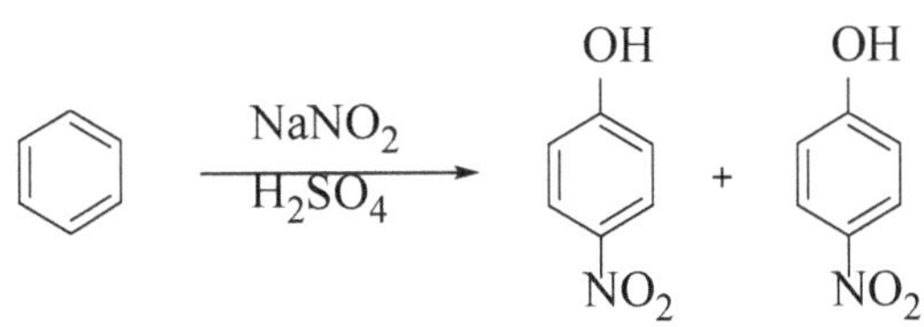

Green approach:

In green method calcium nitrate is solubilized in hot acetic acid and salicylic acid is added to them, after mixing of reaction mixture is heated and poured in ice cold water.

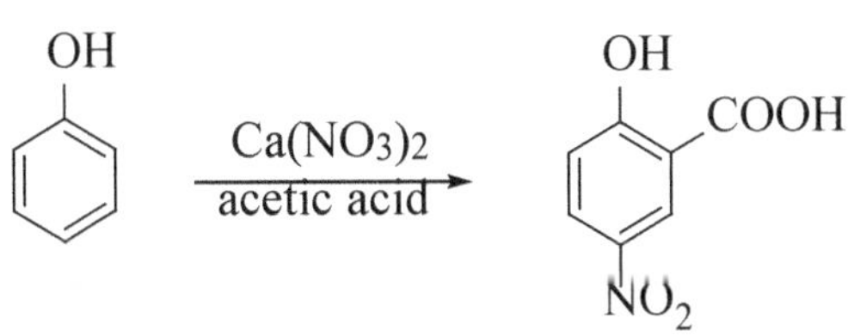

- *Preparation of p-Bromo acetanilide by bromination*

Conventional method:

When acetanilide reacts with bromine in the presence of glacial acetic acid then gives p-bromo acetanilide by bromination reaction.

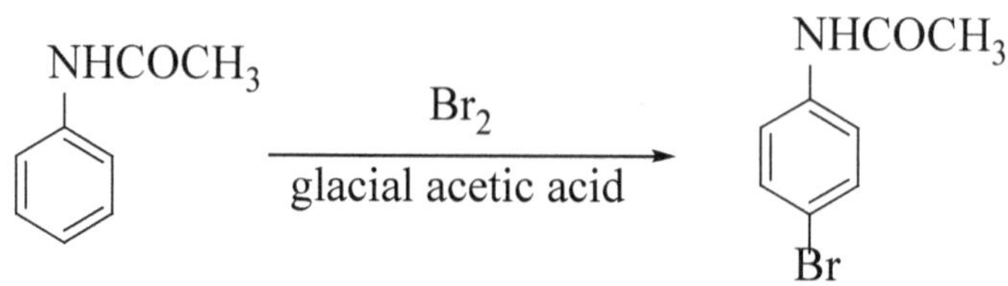

Green method:

Acetanilide is dissolved in ethanol thereafter potassium bromide and ceric ammonium nitrate solution is mix drop wise and stirred for 10 minutes at the room temperature. Now solution is poured into cold water and crystals are filtered.

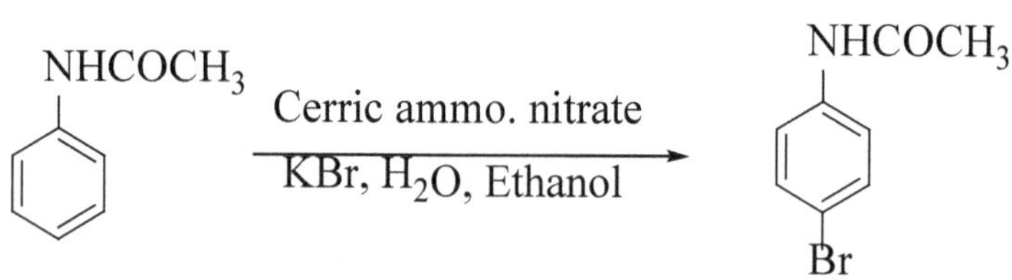

- *Preparation of benzylic acid*

Conventional method:

Benzyl react with KOH in the presence of ethanol to form benzylic acid.

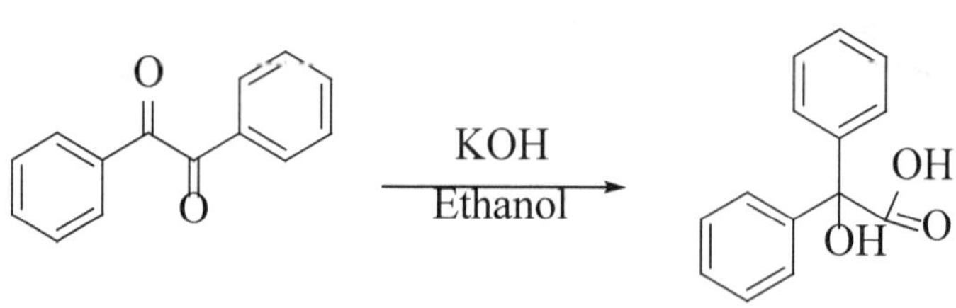

Green approach:

In the green method benzyl and solid KOH are grind in mortar machine which is take hold in dry conical flask and heated on a water bath for 20 minutes. Afterheat it was cool at room temperature, then poured in minimum amount of water and acidified.

9. FUTURE OF GREEN CHEMISTRY

In the coming periods, green chemistry and green nanotechnology continue to growing field to develop sustained environmental management, because influencing scientists, young researcher and engineers globally. The growing global society now adopt teaching and research innovations is more than 25 countries. New many science journals, magazines, several international conferences, seminars and emerging social networking sites for green chemistry has been promote practitioners and collaborators. Many of these collaborations are built around scientist and chemists about the prospective benefits of green chemistry. In continuing impact of green chemistry that way of materials is build, sustainability ideas need to be included all over the educational and teaching process system at globally.

10. CONCLUSION

Recently green chemistry has actively growing field from a tiny idea to a novel broad approach specifically the scientific bases to sustained environmental development and secure. By using green chemistry principles and methods, scientist and researchers can change or modify the traditional methods which are hazardous and not environmentally sustain. Todays, young researchers, chemists and pharma industries demand to be encouraged by considering the principles of green chemistry when designing and selection of reagents and solvents.

Currently green chemistry principles are rapidly transforming industries, leading to the development of ecofriendly products and processes in nanotechnology, pharmaceuticals, agriculture, textiles, cosmetics and electronics. Subsequently sustainable practices are recommending for a life cycle approach, promoting resource efficiency, waste reduction, and responsible end-of-life management through strategies like Reduce, Reuse, Recycle (RRR). Green chemistry and sustainable practices are not mutually exclusive, but rather two sides of the same coin. They work in synergy to create a future where human development and environmental well-being go hand in-hand. Moreover, the favourable outcomes regarding green chemistry is based on the training and skill education system of a coming generation of chemist.

Finally, with due respect the role of sustain education in green chemistry: *"The Biggest Challenge Is to Use Its Rules in Practices of Green Chemistry."*

REFERENCES

1. Anastas, P. T. & Warner, J. C. (1998). Principles of Green Chemistry. Green Chemistry: Theory and Practice. *Oxford University Press, New York.* (ISBN: 9780198506980
2. Anderson, R. (2007). Thematic Content Analysis: *Descriptive Presentation of Qualitative Data. 14*(12).
3. Braun, V. & Clarke, V. (2006). Using Thematic Analysis in Psychology. *Qualitative Research in Psychology, 3*, 77-101.
4. Daryle Busch (Ex-President, ACS) (1998). *Chem. Eng. News, 76*, 47, 12.
5. David, Y. et al., (2000). Environment, Education and Society in the Asia-Pacific, *Routledge Advances in Asia-Pacific Studies*.17:159-161
6. David, T. (2003). Birth to Three Matters: A Review of the Literature: Compiled to Inform the Framework to Support Children in Their Earliest Years. Department for Education and Skills, *Research Report RR444*.
7. Fien, J. (2000). Education for the Environment: A Critique-An Analysis. *Environmental Education Research, 6*, 179-192.
8. Centi, G. & Perathoner, S. (2003). Catalysis and Sustainable (Green) Chemistry. *Catalysis Today, 77*, 4, 287-297.

9. Lozano, R. (2018). Proposing a definition and a framework of organisational sustainability: a review of efforts and a survey of approaches to change. *Sustainability 10*, 1157.
10. Burmeister, F. et al., (2012). Education for Sustainable Development (ESD) and chemistry education, *Chem. Educ. Res. Pract., 13,* 59–68.
11. Eissen, M. & Metzger, J. O. (2002). Environmental Performance Metrics for Daily Use in Synthetic Chemistry. *Chemistry-A European Journal, 8* (16), 3580-3585.
12. Matus, et al., (2007). Overcoming the Challenges to the Implementation of Green Chemistry. CID Working. *Centre for International Development, Harvard University, Cambridge.* 155.
13. Merck Index, *Aldrich Catalogue Chemical Supplier agencies.* N. Thayer, *Cand EN,* 11-19, August 6, 2007.
14. Dunn, P. et al., (2008). Green chemistry tools to influence a medicinal chemistry and research chemistry based organisation. *Green Chemistry, 10,* 31-36.
15. Anastas, P. T. & Williamson, T. C. (1996). Green Chemistry: Designing Chemistry for the Environment. *American Chemical Society, Washington,* DC. ACS Symposium Series Vol. 626 ISBN13: 9780841233997
16. Reports of World Commission on Environment and Development, 1987.
17. Sheldon, R. A. (2014). Green and sustainable manufacture of chemicals from biomass: state of the art. *Green chemistry, 16*(3), 950-963.
18. Roth, A.V. & Menor, L.J. (2003) Insights into Service Operations Management: A Research Agenda. *Production and Operations Management, 12,* 145-164.
19. Tang, S. L. Y. et al., (2005). *Green Chem., 7,* 761.
20. Sheldon, R. A. et al., (2007). *Green chemistry and catalysis.* John Wiley & Sons

21. The Rio Declaration, within *Agenda 21*, 1992.

22. Wardencki, W. et al., (2005). Green Chemistry-Current and Future Issues, *Polish Journal of Environmental Studies, 14* (4), 389-395.

23. Zhang, S. et al., (2017). Precise large deviations of aggregate claims in a size-dependent renewal risk model with stopping time claim-number process, *82*

CHAPTER - 3

THE TOXIC EFFECTS OF ARECOLINE AND ITS IMPLICATIONS

Dr. Barkha Shakya[1*] and Dr. Sonam Shakya[2]

[1]Department of Zoology, Aligarh Muslim University, Aligarh 202002, India

[2]Department of Chemistry, Aligarh Muslim University, Aligarh 202002, India

***Corresponding email: barkhashakya12@yahoo.com**

ABSTRACT

Arecoline, the principal alkaloid found in the areca nut (commonly known as betel nut), is widely consumed across Asia, Africa and the Pacific Islands due to its psychoactive properties and cultural significance. Despite its traditional use, arecoline poses serious health risks, including carcinogenic, neurotoxic, hepatotoxic and cardiovascular effects. As a muscarinic receptor agonist, arecoline mimics acetylcholine, resulting in parasympathetic stimulation that contributes to its stimulant and addictive properties. Chronic exposure has been linked to oral cancer, liver dysfunction, neurodegenerative diseases, and cardiovascular disorders, while acute toxicity can manifest in gastrointestinal distress, bradycardia, and respiratory depression. The addictive nature of arecoline, coupled with its capacity to induce oxidative stress and DNA damage, makes it a public health concern. Additionally, emerging research highlights its potential environmental toxicity. This study underscores the urgent need for public education, regulatory policies and effective treatment strategies to mitigate arecoline-related health and environmental risks, especially in regions where betel nut consumption is culturally ingrained.

Keywords: *Arecoline Toxicity, Carcinogenicity, Muscarinic Receptor Agonist, Oxidative Stress, Public Health Risk, Ecotoxicology*

1. INTRODUCTION

Arecoline, a naturally occurring potent alkaloid found predominantly in the seeds of the areca palm *(Areca catechu)* commonly known as the betel nut. For centuries, Betel nut is chewed, often combined with tobacco, slaked lime and other substances and has been an integral part of cultural and social practices in many regions throughout South and Southeast Asia, Africa, and in parts of the Pacific Islands (Huang et al., 2024; Gunjal et al., 2020). It has been used, both recreationally and in traditional medicine or traditional stimulant (Shah, et al., 2021). Arecoline, the primary bioactive component of betel nut is responsible for its psychoactive and stimulant effects associated with betel nut chewing (Kozlakidis et al., 2022). Though many users enjoy the mild euphoria, increased alertness and pleasure from chewing betel nut, the alkaloid has been linked to numerous adverse health effects when consumed over prolonged periods (Cox et al., 2016). However, its harmful effects on human health and the environment have sparked increasing concern over the years. Beyond its human health impacts, the environmental toxicity of arecoline is also an emerging area of concern (Kozlakidis et al., 2022). This chapter will explore the toxicological effects of arecoline, both on human beings and the broader ecosystem, with a particular focus on its role in ecotoxicology.

2. CHEMICAL PROPERTIES OF ARECOLINE

Arecoline is an alkaloid belonging to the piperidine class of compounds with molecular formula C8H11NO2, consists of a piperidine ring with a carboxyl group attached to a hydroxyl group (Myers, 2022). Arecoline is soluble in water, alcohol and acetone which facilitate its absorption when consumed either through chewing, smoking or swallowing (I. W. G. 2021). The solubility and bioavailability of arecoline enable rapid absorption into the bloodstream, where it can exert its pharmacological effects on various organ systems, particularly the nervous system (Myers, 2022).

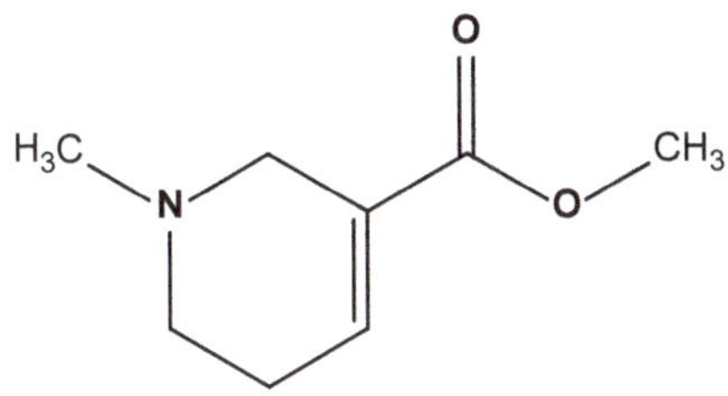

Figure 1: Structure of Arecoline

3. PHARMACOLOGY AND PHARMACOKINETICS OF ARECOLINE

When betel nut is chewed, arecoline is absorbed rapidly through the mucous membranes of the mouth, gastrointestinal tract and respiratory system (Rajarathnam, et al., 2022; Chang, et al., 2014; Forbes, 1964). The rate of absorption of Arecoline can vary depending on the method of consumption, with oral ingestion leading to slower absorption compared to direct mucosal exposure (Forbes, 1964).

Arecoline, a naturally occurring compound with a chemical structure similar to acetylcholine, is classified as a muscarinic receptor agonist (Horenstein et al., 2019; Locock et al., 2016). It predominantly exerts its effects on the central and peripheral nervous systems by mimicking acetylcholine, a neurotransmitter that regulates various physiological processes, including memory, heart rate, and smooth muscle contraction (Wang et al., 2024; Liu et al., 2024). Arecoline binds to muscarinic acetylcholine receptors (mAChRs) found in tissues such as the heart, brain, smooth muscles, and exocrine glands (Liu et al., 2024). By stimulating these receptors, arecoline enhances the parasympathetic activity of the autonomic nervous system, leading to effects such as increased heart rate, salivation and sweating (Garg et al., 2014).

In the brain, arecoline influences several neurotransmitter systems, including dopaminergic, serotonergic and adrenergic systems (Lan et al., 2022; I. W. G., 2021). The release of dopamine contributes to the pleasurable effects experienced by users which underpin the addictive properties of arecoline, though acute effects are relatively mild and transient, chronic exposure to arecoline has been linked to several toxic effects across multiple organ systems (Sun et al., 2024; Liu et al., 2024).

After absorption arecoline is metabolized in the liver by cytochrome P450 enzymes, particularly CYP2A6, which performs N-demethylation to form several metabolites (Lin et al., 2013). Arecoline has a dual role, both as cholinergic agonist and antagonist at specific receptor sites, contributing to its stimulant properties and toxicity (Maher & Reggiani, 2014). The absorption, metabolism, and elimination of arecoline depend on various factors such as dose, frequency of use, and individual metabolic rates (I. W. G., 2021).

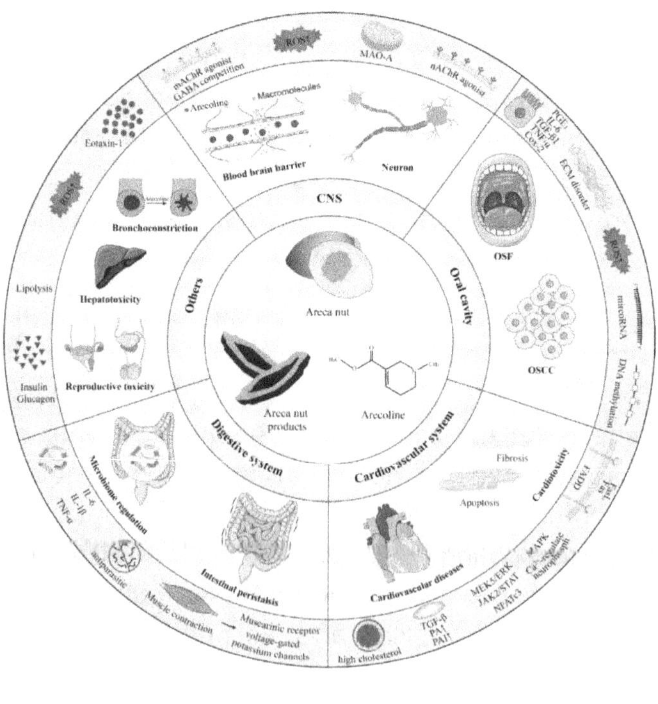

Figure 2: Effects and mechanisms of arecoline on different organs and system (Huang et al., 2024)

4. TOXICOLOGICAL EFFECTS OF ARECOLINE

The health risks associated with arecoline consumption are wide-ranging, spanning from carcinogenicity to neurological and cardiovascular effects (Wang, 2025; Huang et al., 2024). The toxicity of arecoline on human health is increasingly a subject of scientific scrutiny, as chronic exposure to the substance has been linked to severe and sometimes irreversible damage to various organs and systems (Liu et al., 2024).

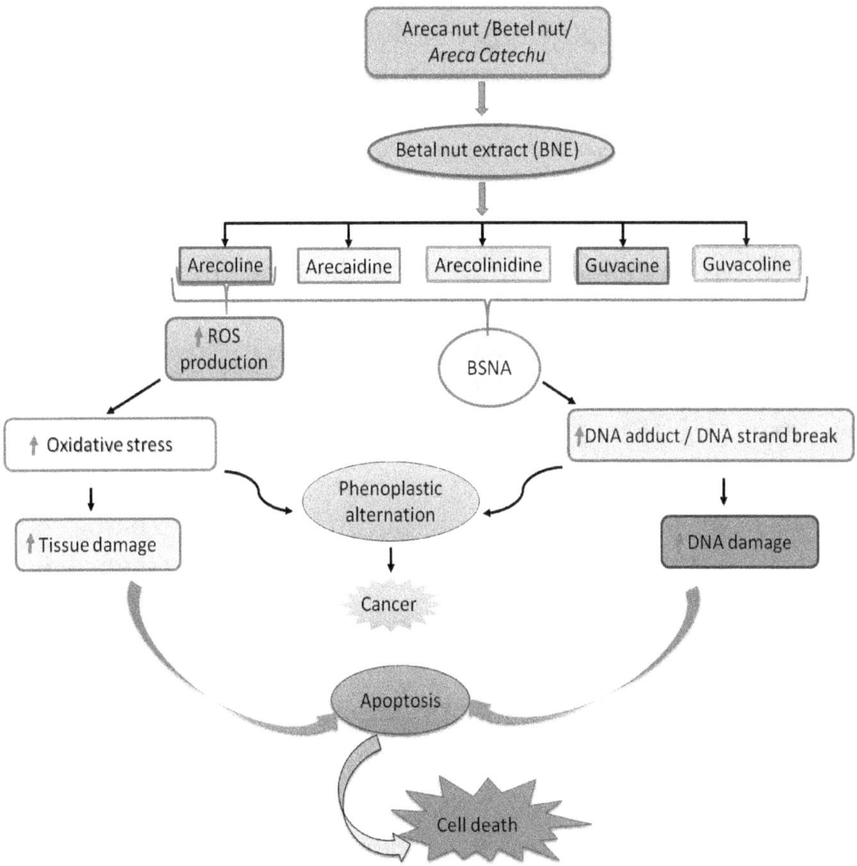

Figure 3: Arecoline's mechanism of action

4.1. Carcinogenicity and Oral Cancer

Arecoline has been extensively studied for its carcinogenic potential, and evidence suggests that long-term use of betel nut products, which contain arecoline, is a major risk factor for oral and other head and neck cancers (Wang et al., 2020).

Areca nut chewing has been classified as a Group 1 carcinogen by the International Agency for Research on Cancer (IARC) (IARC, 2004). This classification is based on sufficient evidence that arecoline and related compounds induce cancer in humans (Senevirathna et al., 2023; Ko et al., 2023). Arecoline's carcinogenic effects are believed to be due to its ability to cause DNA damage, cell mutations, and alterations in cellular pathways (Senevirathna et al., 2023). The compound induces the formation of free radicals, which in turn lead to oxidative stress, a known driver of cancerous transformations in cells (Wang et al., 2024). Moreover, arecoline has been shown to promote the growth of pre-cancerous lesions in the oral cavity, thereby increasing the risk of developing oral squamous cell carcinoma (OSCC) (Srinivasan et al., 2014). This type of cancer is often diagnosed at advanced stages, complicating treatment and prognosis (Bellantoni et al., 2023).

Studies have also pointed to synergistic effects between arecoline and other carcinogens found in betel nut products, such as nitrosamines and tobacco smoke. The combination of these compounds amplifies the carcinogenic risk, making the health consequences of long-term betel nut chewing even more severe (Sharan et al., 2012).

4.2. Neurological Effects and Toxicity

Arecoline exerts significant effects on the central and peripheral nervous systems, primarily due to its parasympathomimetic properties. By mimicking the neurotransmitter acetylcholine, arecoline stimulates muscarinic receptors in the brain and body, which can lead to a range of physiological changes (Wang et al., 2024; Liu et al., 2024). In the short term, arecoline consumption can cause increased salivation, sweating and alterations in heart rate and blood pressure. These effects are generally considered mild and transient at lower doses but can become problematic at higher doses (Liu et al., 2016; Garg et al., 2014; Tan et al., 2014).

Long-term exposure to arecoline, especially through regular betel nut chewing, leads to impaired dopaminergic function in the brain and has been associated with an increased risk of neurodegenerative diseases, such as Parkinson's disease (Huang et al., 2009). Additionally, the addictive nature of arecoline contributes to its long-term consumption and the resulting neurological damage.

The addictive properties are largely attributed to the stimulation of dopamine release, which leads to a sense of euphoria and reinforces continued use (Yan et al., 2024). Dependence on arecoline can develop in individuals who chew betel nut regularly, leading to a cycle of craving and withdrawal that negatively affects mental and emotional health (Liu, et al., 2024; Sumithrarachch et al., 2023).

4.3. Addiction and Behavioral Impact

Betel nut consumption, due to the presence of arecoline, can lead to the development of addiction, which is a significant social and psychological concern (Khatun et al., 2020). Similar to nicotine in tobacco, arecoline triggers the brain's reward system, leading to the dopamine release. This provides users with a temporary sense of pleasure or well-being, but also makes them more likely to develop a psychological dependence on the substance as dependence builds, withdrawal symptoms can include irritability, anxiety, and difficulty concentrating, all of which further encourage continued use. (Huang,et al., 2024; Volgin et al., 2019).

In regions where betel nut chewing is culturally ingrained, the social and behavioral effects of addiction are substantial (Huang et al., 2017). Addicted individuals may experience a decrease in quality of life, as their daily activities are often centered around the need to obtain and consume arecoline-containing products (Ghosh et al., 2017). In severe cases, the addiction can affect family dynamics and lead to social stigma.

4.4. Hepatotoxicity

Arecoline is metabolized by the liver, and chronic exposure to high doses can lead to liver toxicity, manifesting as liver dysfunction or damage (Wang, et al., 2017). While the liver efficiently metabolizes many substances, the processing of arecoline can induce oxidative stress, inflammation, and cellular injury (Run-mei, et al., 2014). Research has shown that arecoline can lead to the accumulation of reactive oxygen species (ROS) in hepatocytes, causing mitochondrial damage and activation of apoptotic pathways (Wang, et al., 2018).

The hepatotoxic effects of arecoline are believed to stem from the generation of ROS, which can damage liver cells, disrupt mitochondrial function, and promote inflammation (Sun, et al., 2018). Chronic exposure to these oxidative stressors may impair liver function, resulting in elevated liver enzymes such as ALT and AST, which are indicative of liver injury (Run-mei, et al., 2014). Over time, this damage could lead to conditions such as fatty liver disease, fibrosis, or even cirrhosis (Liu, et al., 2024; Zhu, et al., 2023; Sun, et al., 2018).

4.5. Cardiovascular Effects

Chronic arecoline consumption has been associated with several cardiovascular effects, including increased heart rate and blood pressure (Qiang, 2023). Arecoline stimulates the parasympathetic nervous system, which regulates the body's automatic functions, such as heart rate and digestion (Tong et al., 2024; Kozlakidis et al., 2022) At low doses, these effects might be minimal and transient; however, prolonged use can lead to elevated blood pressure and heart rate, which places undue stress on the cardiovascular system (Ho et al., 2022). Arecoline can worsen and exacerbate these cardiovascular issues in people with conditions like hypertension or heart disease, increasing the risk of heart attacks, strokes, and other related events (Choudhury et al., 2012).

Furthermore, the presence of other toxic compounds in betel nut preparations, such as nicotine and polycyclic aromatic hydrocarbons (PAHs), can compound the cardiovascular risks associated with arecoline use (Noor et al., 2024).

5. CLINICAL MANIFESTATIONS OF ARECOLINE TOXICITY

Arecoline toxicity can manifest in a wide array of symptoms, which can be divided into acute and chronic toxicity (Gao et al., 2024; Serikuly et al., 2021).

5.1. Acute Toxicity

Acute toxicity typically occurs when large amounts of arecoline or betel nut are consumed in a short period of time (Wang et al., 2024). Common symptoms of acute poisoning include:

- **Nausea and vomiting**: Often the first signs of toxicity, caused by the stimulation of muscarinic receptors in the gastrointestinal tract (Gilani et al., 2024; Mortazavi et al., 2023).
- **Excessive salivation (sialorrhea)**: An exaggerated response due to overstimulation of salivary glands (Agustina et al., 2024; Garg et al., 2023).
- **Abdominal pain and diarrhea**: Resulting from gastrointestinal irritation and increased peristalsis (Mortazavi et al., 2023).
- **Bradycardia**: A reduction in heart rate due to muscarinic receptor activation in the heart (Lohith et al., 2013; Sun et al., 2010).
- **Hypotension**: Low blood pressure, which can lead to dizziness, fainting, and in severe cases, circulatory shock (Liu et al., 2024; Deng et al., 2001).
- **Respiratory depression**: A life-threatening consequence that can occur when the cholinergic system is overstimulated, leading to reduced respiratory drive (Perman et al., 2021).

In severe cases, patients may experience muscle weakness, seizures and coma as a result of widespread cholinergic dysfunction (Garg et al., 2014; Chang et al., 2013; Wu et al., 2012;

5.2. Chronic Toxicity

Chronic toxicity results from prolonged or repeated exposure to arecoline and betel nut, and it manifests as:

- **Oral and Esophageal Cancers**: Chronic irritation and DNA damage in the oral and esophageal mucosa can lead to the development of cancers, particularly squamous cell carcinoma (Choudhury et al., 2025; Li et al., 2022).

- **Neurodegenerative Diseases**: Long-term exposure to arecoline has been linked to cognitive impairment and neurodegenerative disorders such as Alzheimer's and Parkinson's disease. The cumulative effects of oxidative stress and cholinergic disruption contribute to these conditions (Yan et al., 2024).
- **Liver Damage**: Arecoline-induced oxidative stress can lead to hepatotoxicity, with signs such as hepatomegaly (enlarged liver), jaundice, and in extreme cases, cirrhosis (Sun et al., 2023).
- **Addiction and Dependence**: Betel nut, due to the presence of arecoline, has addictive properties. Chronic users may experience withdrawal symptoms such as anxiety, irritability, and difficulty concentrating upon cessation (Li et al., 2025; Papke et al., 2015).

6. PUBLIC HEALTH IMPLICATIONS

Arecoline-induced toxicity represents a significant public health challenge, especially in regions where betel nut consumption is widespread. The adverse effects associated with arecoline, including its carcinogenic, neurotoxic, hepatotoxic and addictive properties, highlight the need for public health interventions aimed at reducing consumption and raising awareness about its potential health risks (Kozlakidis et al., 2022; Chatterjee & Gupte., 2023).

6.1. Prevention and Education

Public health campaigns should focus on educating communities about the dangers of betel nut chewing, emphasizing the risks of oral and esophageal cancers, liver disease, and addiction. Education campaigns should also highlight the role of arecoline in these adverse health outcomes and provide information on healthier alternatives (Dhingra et al., 2023; Das et al., 2020).

6.2. Regulation and Control

Some countries have implemented regulations restricting the sale and consumption of betel nut, but these efforts are often insufficient or poorly enforced (Croucher, & Islam 2002).

Stricter regulations, such as age restrictions, labeling requirements, and public health warnings, could help reduce the prevalence of betel nut chewing and minimize the associated health risks (Chatterjee & Gupte., 2023; Farrand et al., 2001).

6.3. Treatment for Dependence

As arecoline has addictive properties, treatment strategies for individuals with betel nut dependence should be explored. Behavioral therapies, such as cognitive-behavioral therapy (CBT), and pharmacological interventions could be employed to help individuals reduce their dependence on betel nut (Herzog et al., 2023; Athukorala et al., 2021; Hussain et al., 2018).

7. CONCLUSION

Arecoline-induced toxicity is a significant health concern, particularly in regions where betel nut chewing is prevalent. Arecoline is a potent toxic substance with significant implications for both human health and the environment. The toxicity of arecoline arises from multiple mechanisms, including cholinergic overstimulation, oxidative stress, neurotoxicity, carcinogenesis and addictive properties pose serious risks to individuals who regularly consume betel nut products. Acute toxicity can lead to life-threatening symptoms, while chronic exposure can result in severe long-term health issues such as cancer, liver damage, and neurodegeneration.

Arecoline, a bioactive alkaloid from the areca nut, poses emerging concerns in ecotoxicology due to its persistence and toxicity in aquatic and terrestrial environments. As a water-soluble compound, arecoline can leach into soil and waterways from improperly discarded betel nut waste, affecting microbial communities, aquatic invertebrates, and plant systems. Its ability to bioaccumulate raises concerns about biomagnification through food chains, potentially harming higher organisms. Studies suggest arecoline disrupts enzymatic activity, induces oxidative stress, and affects reproduction in aquatic species. Given its widespread use and improper disposal, understanding arecoline's environmental fate and toxicological effects is essential for ecosystem protection and sustainability.

As research continues to uncover the full extent of arecoline's toxicity, it is critical to develop regulatory frameworks and mitigation strategies to protect both human and environmental health.Only by understanding the broader implications of arecoline use can we hope to address the risks associated with its widespread consumption and ecological impact. Education about the harmful effects of arecoline and betel nut, along with better prevention and treatment strategies, is essential for reducing the burden of arecoline-induced toxicity.

REFERENCES

1. Agustina, R. et al., (2024). Betel nut chewing effects on the incidence of periodontitis. *World Journal of Advanced Research and Reviews, 23*(3).
2. Athukorala, I. A. et al., (2021). Areca nut chewing: initiation, addiction, and harmful effects emphasizing the barriers and importance of cessation. *Journal of addiction, (1)*, 9967097.
3. Awang, M. N. (1986). Estimation of arecoline contents in commercial areca (betel) nuts and its relation to oral precancerous lesions. *Singapore Med J, 27*(4), 317-20.
4. Chang, Y. F. et al., (2013). Arecoline inhibits and destabilizes agrin-induced acetylcholine receptor cluster formation in C2C12 myotubes. *Food and chemical toxicology, 60*, 391-396.
5. Chatterjee, N. & Gupte, H. A. (2023). Areca nut use among adolescents: How do we prevent and control this problem?*Journal of Global Health, 13*, 03022.
6. Choudhury, M. D. et al., (2012). Atherogenic effect of Arecoline: A computational study. *Bioinformation, 8*(5), 229.

7. Choudhury, Y. et al., (2025). A review on repurposing anti-diabetic drugs for the amelioration of betel-nut induced carcinogenesis. *Indian Journal of Biochemistry and Biophysics (IJBB)*, *62*(3), 222-236.

8. Croucher, R. & Islam, S. (2002). Socio-economic aspects of areca nut use. *Addiction biology*, *7*(1), 139-146.

9. Das, A. et al., (2020). Areca nut and betel quid control interventions: halting the epidemic. *Substance Use & Misuse*, *55*(9), 1552-1559.

10. Deng, J. F. et al., (2001). Acute toxicities of betel nut: rare but probably overlooked events. *Journal of Toxicology: Clinical Toxicology*, *39*(4), 355-360.

11. Dhingra, K. & Jhanjee, S. (2023). A review of intervention strategies for areca nut use cessation. *Indian Journal of Psychological Medicine*, *45*(2), 117-123.

12. Farrand, P. et al, (2001). Prevalence, age of onset and demographic relationships of different areca nut habits amongst children in Tower Hamlets, London. *British dental journal*, *190*(3), 150-154.

13. Forbes, L. S. (1964). The use of arecoline hydrobromide in the treatment of taeniid infections in dogs. *Annals of Tropical Medicine & Parasitology*, *58*(1), 116-118.

14. Gao, W. et al., (2024). Comprehensive Insights into Arecoline Hydrobromide: Pharmacology, Toxicity, and Pharmacokinetics. *Medical Science Monitor: International Medical Journal of Experimental and Clinical Research*, *30*, e945582.

15. Garg, A. et al., (2014). A review of the systemic adverse effects of areca nut or betel nut. *Indian journal of medical and paediatric oncology*, *35*(01), 3-9.

16. Ghosh, S. et al., (2021). JunD accentuates arecoline-induced disruption of tight junctions and promotes epithelial-to-mesenchymal transition by association with NEAT1 lncRNA. *Oncotarget*, *12*(15), 1520.

17. Gilani, A. H. et al., (2004). Presence of cholinomimetic and acetylcholinesterase inhibitory constituents in betel nut. *Life Sciences, 75*(20), 2377-2389.

18. Giri, S. et al., (2010). Arecoline induced disruption of expression and localization of the tight junctional protein ZO-1 is dependent on the HER 2 expression in human endometrial Ishikawa cells. *BMC Cell Biology, 11*, 1-12.

19. Herzog, T. A. et al., (2023). The Betel Nut Intervention Trial (BENIT)—A randomized clinical trial for areca nut and betel quid cessation: primary outcomes. *International Journal of Environmental Research and Public Health, 20*(16), 6622.

20. Ho, T. J. et al., (2022). Arecoline induces cardiotoxicity by upregulating and activating cardiac hypertrophy-related pathways in Sprague–Dawley rats. *Chemico-biological interactions, 354*, 109810.

21. Huang, G. et al., (2024). Recent Advance on Biological Activity and Toxicity of Arecoline in Edible Areca (Betel) Nut: A Review. *Foods, 13*(23), 3825.

22. Hussain, A. et al., (2018). Reasons for betel quid chewing amongst dependent and non-dependent betel quid chewing adolescents: a school-based cross-sectional survey. *Substance Abuse Treatment, Prevention, and Policy, 13*, 1-8.

23. Bellantoni, I. et al., (2023). Oral cavity squamous cell carcinoma: an update of the pharmacological treatment. *Biomedicines, 11*(4), 1112.

24. Khatun, J. et al., (2020). Betel nut addiction detection using machine learning. In *2020 4th International Conference on Electronics, Communication and Aerospace Technology (ICECA)* (pp. 1435-1440). IEEE.

25. Ko, A. M. S. et al., (2023). Systematic review of roles of arecoline and arecoline N-oxide in oral cancer and strategies to block carcinogenesis. *Cells, 12*(8), 1208.

26. Kozlakidis, Z. et al., (2022). Betel nut and arecoline: past, present, and future trends. *Innovations in Digital Health, Diagnostics, and Biomarkers, 2*(2022), 64-72.

27. Li, J. et al., (2022). Arecoline is associated with inhibition of cuproptosis and proliferation of cancer-associated fibroblasts in oral squamous cell carcinoma: a potential mechanism for tumor metastasis. *Frontiers in oncology, 12*, 925743.

28. Li, X. et al., (2025). Modulation of Neural Activity and Gene Expression by Arecoline. *Frontiers in Integrative Neuroscience, 19*, 154526

29. Lin, C. C. et al., (2017). Social and cultural context of betel quid consumption in Taiwan and implications for prevention and cessation interventions. *Substance Use & Misuse, 52*(5), 646-655.

30. Liu, H. et al., (2024). Review of the toxic effects and health functions of arecoline on multiple organ systems. *Food Innovation and Advances, 3*(1), 31-41.

31. Liu, Y. J. et al., (2016). The pharmacology, toxicology and potential applications of arecoline: a review. *Pharmaceutical Biology, 54*(11), 2753-2760.

32. Lohith, T. S. et al., (2013). Acute oral toxicity of raw areca nut extract in rats. *Indian Journal of Animal Research, 47*(5), 431-434.

33. Mortazavi, M. et al., (2023). Acute toxic effects of areca nut on central nervous system and liver: A case report. *Clinical Case Reports, 11*(10), e7976.

34. Myers, A. L. (2022). Metabolism of the areca alkaloids–toxic and psychoactive constituents of the areca (betel) nut. *Drug Metabolism Reviews, 54*(4), 343-360.

35. Noor, A. E. & Ramanarayana, B. (2024). Relationship of smokeless tobacco uses in the perspective of oral cancer: A global burden. *Oral Oncology Reports*, 100516.

36. Papke, R. L. et al., (2015). Nicotinic activity of arecoline, the psychoactive element of" betel nuts", suggests a basis for habitual use and anti-inflammatory activity. *PloS one, 10*(10), e0140907.

37. Perman, B. et al., (2021). Acute eosinophilic pneumonia following areca nut use. *Chest, 160*(4), A1685.

38. Qiang, X. 2023. Exploring the Potential Medicinal Uses of Arecoline and a Comprehensive Overview of its Effects on Human Health. Der Pharma Lett.15:11-12.

39. Run-mei, X. et al., (2014). Effects of arecoline on hepatic cytochrome P450 activity and oxidative stress. *The Journal of Toxicological Sciences, 39*(4), 609-614.

40. Senevirathna, K. et al., (2023). Carcinogenic effects of areca nut and its metabolites: a review of the experimental evidence. *Clinics and practice, 13*(2), 326-346.

41. Serikuly, N. et al., (2021). Effects of acute and chronic arecoline in adult zebrafish: Anxiolytic-like activity, elevated brain monoamines and the potential role of microglia. *Progress in Neuro-Psychopharmacology and Biological Psychiatry, 104*, 109977.

42. Sharan, R. N. et al., (2012). Association of betel nut with carcinogenesis: revisit with a clinical perspective.

43. Sumithrarachchi, S. & Jayasinghe, R. (2023). Pharmaceutical agents for areca nut cessation. In *Oral Submucous Fibrosis: A Guide to Diagnosis and Management* (pp. 349-355). Cham: Springer International Publishing.

44. Sun, J. et al., (2023). Arecoline-induced hepatotoxicity in rats: screening of abnormal metabolic markers and potential mechanisms. *Toxics, 11*(12), 984.

45. Sun, Y. P. et al., (2010). Systemic administration of arecoline reduces ethanol-induced sleeping through activation of central muscarinic receptor in mice. *Alcoholism: Clinical and Experimental Research, 34*(1), 150-157.

46. Tong, T. et al., (2024). Biological effects and biomedical applications of areca nut and its extract. *Pharmaceuticals, 17*(2), 228.

47. Volgin, A. D. et al., (2019). DARK classics in chemical neuroscience: arecoline. *ACS Chemical Neuroscience, 10*(5), 2176-2185.

48. Wang, T. H. et al., (2024). Arecoline Induces ROS Accumulation, Transcription of Proinflammatory Factors, and Expression of KRT6 in Oral Epithelial Cells. *Biomedicines*, *12*(2), 412.

49. Wang, T. S. et al., (2018). CYP450-mediated mitochondrial ROS production involved in arecoline N-oxide-induced oxidative damage in liver cell lines. *Environmental Toxicology*, *33*(10), 1029-1038.

50. Wang, W. (2025). Assessing the carcinogenic potential and molecular mechanisms of arecoline in human lungs: from in silico methods to in vitro validation. *Ecotoxicology and Environmental Safety*, *294*, 118108.

51. Wang, X. et al., (2018). Effect of autophagy-associated proteins on the arecoline-induced liver injury in mice. *Experimental and Therapeutic Medicine*, *16*(4), 3041-3049.

52. Wang, Y. C. et al., (2020). Downregulation of the DNA repair gene DDB2 by arecoline is through p53's DNA-binding domain and is correlated with poor outcome of head and neck cancer patients with betel quid consumption. *Cancers*, *12*(8), 2053.

53. Wang, Z. et al., (2024). A review of the traditional uses, pharmacology, and toxicology of areca nut. *Phytomedicine*, 156005.

54. Wu, M. et al., (2012). Four types of traditional Chinese medicine inducing epileptic seizures. *Seizure*, *21*(5), 311-315.

55. Yan, H. et al., (2024). Arecoline induces neurotoxicity in HT22 cells via the promotion of endoplasmic reticulum stress and downregulation of the Nrf2/HO-1 pathway. *Environmental Toxicology*, *39*(6), 3410-3424.

56. Zhu, L. et al., (2023). Gut metabolomics and 16S rRNA sequencing analysis of the effects of arecoline on non-alcoholic fatty liver disease in rats. *Frontiers in Pharmacology*, *14*, 1132026.

Chapter - 4

ENVIRONMENTAL CONTAMINATION: AN INTEGRATED ANALYSIS OF GLOBAL POLLUTION

Deeksha Singh[1] and Dr. Sonal Singh[2*]

[1]Department of Zoology, Agra College, Agra, Affiliated to Dr. Bhimrao Ambedkar University, Agra

[2]Department of Zoology, D.S. College, Aligarh, Affiliated to Raja Mahendra Pratap Singh State University, Aligarh

Corresponding email: drsonal.aca@gmail.com

ABSTRACT

This chapter provides a thorough overview of the deep and interrelated effects that pollution of air, water, and soil has on ecosystems and human health around the world. Pollution, caused by swift industrialisation, urbanisation, and unsustainable human practices, constitutes a significant threat that overcomes geographical limitations and media outlets. The overview summarises research on the main causes and ways that pollution happens, focussing on important pollutants such particulate matter, nitrogen oxides, heavy metals, persistent organic pollutants, and microplastics. It looks at how air pollution affects people and the environment in a chain reaction, causing respiratory and cardiovascular disorders in people and acid rain and smog that alter ecosystems. It also talks about the effects of water pollution, such as the spread of illnesses that live in water, the loss of aquatic biodiversity, and the breakdown of food chains. Lastly, the chapter talks on soil degradation and how pollution hurts farming, poisons food sources, and stays in the soil for a long time, where it can seep into groundwater. This chapter contends that a comprehensive approach to environmental management is necessary, considering the fundamental connection between ecosystem health and human well-being through the combination of various scientific viewpoints.

Keywords: *Environmental pollution, Ecosystem Health, air pollution, water contamination, and soil degradation*

1. INTRODUCTION

The relationship between people and the natural world has reached a critical point in a time of rapid technological progress and population growth. Modern civilisation has brought many benefits, such as more industrial manufacturing and more trade between countries. However, these benefits have come at a great cost to the environment, with pollution being one of the biggest and most complex threats to life on Earth. Pollution is generally defined as the introduction of harmful substances into the natural environment that cause changes for the worse. These pollutants can be chemicals, energy, or living things, and they can come from either point sources (like factory chimneys or sewage outlets) or non-point sources (like stormwater or runoff from farms). People used to think of pollution as separate things: air, water, and soil. But new research shows that things are more complicated than they seem. These environmental media are not separate; they are part of a complex system that works together. Pollutants that get into the air can land on land and water, and those that get into the soil can leach into groundwater or become airborne through wind erosion. This chapter examines the individual and collective effects of air, water, and soil pollution on ecosystems and human health, utilising contemporary research to underscore the pressing necessity for integrated, systemic solutions. This analysis is based on the idea that the health of people is directly related to the health of the environment, and that a healthy planet is necessary for people to be satisfied.

2. THE IMPACT OF AIR POLLUTION

Air pollution is a major environmental health risk, responsible for millions of premature deaths globally each year (Fuller et al., 2022). It is a complex mixture of gaseous and particulate matter that can travel vast distances, affecting populations far from the source. The primary sources include industrial processes, energy production, vehicular emissions, and the burning of biomass.

2.1. Key Air Pollutants and Their Sources

I. **Particulate Matter** (PM)**:** Fine inhalable particles with a diameter of 2.5 micrometers or less (PM 2.5) pose the greatest health risk. They are produced by combustion processes and can penetrate deep into the lungs and even enter the bloodstream (WHO, 2021).

II. **Nitrogen Oxides** (NOx) **and Sulfur Dioxide** (SO2)**:** These gases are primarily released from fossil fuel combustion in vehicles and power plants. They are key precursors to acid rain and the formation of secondary pollutants like ozone and aerosols.

III. **Ground-Level Ozone** (O3)**:** Unlike the stratospheric ozone layer which protects against UV radiation, ground-level ozone is a harmful pollutant formed by the reaction of NOx and volatile organic compounds (VOCs) in the presence of sunlight.

IV. **Carbon Monoxide** (CO)**:** A toxic gas produced by the incomplete combustion of carbon-based fuels.

2.2. Effects on Human Health

The human body is highly susceptible to the negative effects of air pollution. Research has established a clear link between exposure to air pollutants and a wide range of health issues. Long-term exposure to PM2.5 is a leading cause of cardiovascular disease, lung cancer, and chronic obstructive pulmonary disease (COPD) (Lelieveld *et al.,* 2015). The small size of these particles allows them to bypass the body's natural defenses, causing inflammation and oxidative stress throughout the body. NOx and SO2 can irritate the airways and exacerbate conditions like asthma. Studies have also linked air pollution to adverse birth outcomes, neurological disorders, and mental health issues (Fuller *et al.*, 2022).

2.3. Effects on Ecosystems

Air pollution's impact extends far beyond human health, fundamentally altering natural ecosystems.

1. **Acid Rain:** The deposition of sulfur and nitrogen compounds from the atmosphere into the environment via precipitation. This process lowers the pH of soil and water bodies, which can have devastating consequences. In aquatic ecosystems, it can lead to the death of fish and other organisms, while in forests, it can damage trees and impair nutrient uptake from the soil.
2. **Smog and Reduced Visibility:** The hazy, brownish smog that blankets many urban areas is a visual manifestation of air pollution. Smog can stunt plant growth and reduce agricultural yields by damaging leaves and interfering with photosynthesis.
3. **Climate Change:** Many air pollutants, such as black carbon (a component of PM), are powerful short-lived climate pollutants that contribute to global warming. Reducing these pollutants can yield both air quality and climate benefits (UNEP, 2021).

3. THE IMPACT OF WATER POLLUTION

Clean, safe water is a basic human right and an important part of healthy ecosystems. For billions of people and countless species, though, water contamination makes this resource less useful. It happens when industrial waste, agricultural runoff, municipal sewage, and rainwater from cities are dumped into lakes, rivers, oceans, and groundwater.

3.1 Key Water Pollutants and Their Sources

I. **Industrial Effluents:** Heavy metals (e.g., lead, mercury, arsenic), organic chemicals, and other toxic substances are often discharged from industrial facilities, directly poisoning aquatic life and rendering water unsafe for human consumption.

II. **Agricultural Runoff:** The widespread use of fertilizers and pesticides leads to the runoff of nitrates, phosphates, and agrochemicals into waterways.

III. **Wastewater and Sewage:** Untreated or inadequately treated sewage introduces pathogens (bacteria, viruses) and nutrients into water bodies.

IV. **Plastics and Microplastics:** The ubiquitous presence of plastic waste in aquatic environments poses a severe threat, from entanglement of marine animals to the ingestion of microplastics, which can carry toxins (Fuller *et al.*, 2022).

3.2. Effects on Human Well-being

The health consequences of water pollution for humans are profound. Contaminated drinking water is a primary vector for waterborne diseases such as cholera, typhoid, and giardiasis. According to the World Health Organization, contaminated drinking water is responsible for millions of deaths annually, particularly among young children (WHO, 2021). In addition to pathogens, chemical contaminants in water can lead to chronic diseases. Exposure to heavy metals like lead can cause neurological damage, while persistent organic pollutants can disrupt the endocrine system and increase cancer risk.

3.3. Effects on Ecosystems

Water pollution devastates aquatic ecosystems. The introduction of excess nutrients from agricultural runoff and sewage leads to a process known as **eutrophication**. This over-enrichment causes rapid algal blooms, which block sunlight, consume dissolved oxygen, and create "dead zones" where most aquatic life cannot survive. Heavy metals and other toxins accumulate in the food chain through a process called **biomagnification**, reaching dangerously high concentrations in top predators, including humans. The plastic debris crisis is also a major concern, as it not only harms wildlife but also disrupts delicate marine habitats.

4. THE IMPACT OF SOIL POLLUTION

Soil is the foundation of terrestrial life, providing the medium for plant growth, filtering water, and supporting a vast diversity of organisms. However, soil pollution, often a silent and invisible threat, undermines these vital functions. It is caused by industrial activities, agricultural practices, mining, and improper waste disposal (Zhu *et al.*, 2022).

4.1 Key Soil Pollutants and Their Sources

I. **Heavy Metals:** Lead, mercury, cadmium, and arsenic are common soil contaminants from industrial waste, mining activities, and the use of certain pesticides (Zhu *et al.*, 2022).

II. **Pesticides and Herbicides:** The chemicals used in agriculture to control pests and weeds can persist in the soil for decades, contaminating groundwater and affecting soil microbes.

III. **Industrial Chemicals:** Spills and improper disposal of chemicals like petroleum hydrocarbons, polycyclic aromatic hydrocarbons (PAHs), and polychlorinated biphenyls (PCBs) can render large areas of soil unusable (Wu and Zhang, 2020).

IV. **Solid Waste:** Landfills and illegal dumping sites introduce a mixture of pollutants into the soil, including plastics, metals, and chemicals that can leach into the surrounding environment.

4.2. Effects on Human Well-being

The link between soil pollution and human health is multifaceted. First, contaminated soil compromises food security and safety. Pollutants like heavy metals can be taken up by food crops, entering the food chain and posing a direct threat to human health (Zhu et al., 2022). For example, cadmium contamination in rice has been a serious public health issue in parts of Asia. Second, direct exposure to polluted soil can occur through contact, inhalation of dust, or ingestion, particularly for children playing in contaminated areas. These exposures can lead to a variety of health problems, from developmental issues to cancer.

4.3. Effects on Ecosystems

Soil pollution directly impacts ecosystem function. It can destroy the intricate microbial communities that are essential for nutrient cycling and soil health, leading to a loss of fertility. Contaminated soil can also harm plants, stunting their growth or causing their death. The pollutants in the soil act as a long-term sink, from which they can be re-released into other environmental media. For instance, wind erosion can make contaminated soil particles airborne, and rainfall can cause pollutants to leach into rivers and aquifers.

Table 1: A Comparison of Key Pollutants, Sources, and Their Impacts

Pollutant Category	Primary Sources	Impact on Ecosystems	Impact on Human Well-being
Air	Industrial emissions, vehicles, power plants, biomass burning	Acid rain, smog, climate change, reduced crop yields	Respiratory diseases, cardiovascular issues, cancer
Water	Industrial discharge, agriculture, sewage, plastic waste	Eutrophication, dead zones, biomagnification, loss of biodiversity	Waterborne diseases, neurological disorders, cancer risk
Soil	Industrial waste, pesticides, improper solid waste disposal	Loss of soil fertility, destruction of microbial communities, contamination of food crops	Contamination of food chain, developmental issues, cancer

CONCLUSION

Pollution of the air, water, and soil is a serious and interrelated problem that threatens the health of our planet's ecosystems and our own survival, according to the studies discussed in this chapter. The traditional approach of addressing these issues in isolation is insufficient. Pollutants discharged into one medium eventually end up in another, setting off a chain reaction of complicated and far-reaching health and environmental issues. An integrated, comprehensive strategy to environmental management is necessary to successfully tackle this worldwide challenge. A combination of stronger regulatory frameworks, funding for clean technology, and a shift in social consumption habits is required to address this. In addition, encouraging people to take responsibility for the environment requires recommitting resources to public education and awareness campaigns. The well-being of our communities and the health of our ecosystems are interdependent issues. We can start to repair the damage that has already been done and create a better and more sustainable future for everyone if we acknowledge this deep connection and act swiftly on the findings of scientific research.

REFERENCES

1. Fuller, R. et al., (2022). Pollution and health: a progress update. *The Lancet Planetary Health, 6*(6), e535-e547.
2. Lelieveld, J. et al., (2015). The contribution of outdoor air pollution sources to premature mortality on a global scale. *Nature, 525*(7568), 367-371.
3. United Nations Environment Programme (UNEP). (2021). *UNEP Frontiers 2021 Report: From a Human-centered Planet to a Nature-based Planet*. Nairobi: UNEP.
4. World Health Organization. (2021). *WHO global air quality guidelines: Particulate matter (*PM2.5 *and* PM10*), ozone, nitrogen dioxide, sulfur dioxide and carbon monoxide*. Geneva: World Health Organization.

5. Wu, S. & Zhang, Y. (2020). The inter-linkage of pollution and human health: A systematic literature review. *Journal of Environmental Management, 270,* 110900.

6. Zhu, M. et al., (2022). The impacts of soil pollution on human health and ecosystem: A comprehensive review. *Science of The Total Environment, 806,* 150650.

Chapter - 5

N-NITROSODIETHYLAMINE (NDEA): ENVIRONMENTAL PATHWAYS, HUMAN EXPOSURE AND TOXICOLOGICAL IMPACT

Harikesh Kumar Singh[1], Zubaida Khatoon[1], Dr. Shalini Singh[2] and Dr. Hadiya Husain[1*]

[1]Department of Zoology, Faculty of Science, University of Lucknow, Lucknow-226007

[2]Government Degree College Kuchlai, Sitapur-261303

***Corresponding author: hadiya.husain@gmail.com**

ABSTRACT

One volatile nitrosamine known for its strong carcinogenic effects is **N-nitrosodiethylamine (NDEA)**. The International Agency for Research on Cancer (IARC) has classified NDEA as a probable human carcinogen (Group 2A). It is detected in various environmental matrices, as it is generated through industrial processes, tobacco smoke and as a by-product of several manufacturing activities. The presence of NDEA in contaminated food and drinking water has raised significant public health concerns, as numerous studies have reported its occurrence in a wide range of food products and water sources. Once NDEA enters the human body, its metabolic activation is primarily mediated by cytochrome P450 enzymes, particularly CYP2E1. This bioactivation results in the formation of highly reactive electrophilic intermediates that can bind to DNA, leading to DNA adduct formation and initiation of mutagenic processes that may ultimately result in carcinogenesis. Animal studies have demonstrated that NDEA induces tumors in multiple organs, including the liver, lungs, and esophagus. Owing to its widespread environmental occurrence and severe health risks, stringent regulatory measures, advanced detection techniques, and increased public awareness are urgently required to minimize NDEA exposure and its associated toxicological effects.

Keywords: *NDEA, environmental pathways, carcinogenic, human health, toxicology*

1. INTRODUCTION

The purpose of this chapter is to investigate the effects of **N-nitrosodiethylamine (NDEA)** on human health arising from exposure through different environmental and anthropogenic sources. The adverse impacts of environmental pollution have intensified with rapid economic development, particularly in developing countries such as China, India and Iran. Increased industrialization, extensive use of industrial and non-industrial machinery, emissions from fossil fuel–based industries, and reduced rainfall have collectively contributed to the release of toxic gases and pollutants into the environment (Fehresti-Sani et al., 2024).

The International Agency for Research on Cancer (IARC) has classified N-nitrosodiethylamine (NDEA), a volatile and water-soluble nitrosamine compound, as a probable human carcinogen (Group 2A) (IARC, 2010). NDEA belongs to the N-nitrosamine family, which has received considerable attention in toxicological research due to its strong mutagenic, genotoxic and carcinogenic properties (Lijinsky, 1992). Unlike many industrial chemicals, NDEA is not intentionally manufactured for commercial purposes; rather, it is formed as a by-product of various industrial, environmental and biological processes.

Environmental contamination with NDEA can occur during the manufacture and degradation of rubber products, pesticides, cosmetics, pharmaceuticals and through industrial effluents. Additionally, NDEA may form during water disinfection processes, particularly chlorination of water containing secondary amines (Mitch & Sedlak, 2002). Currently, nine different nitrosamines are recognized as potent carcinogens and are classified as emerging nitrogen-containing disinfection by-products (N-DBPs). Although most studies have focused on N-nitrosodimethylamine (NDMA) due to its frequent detection and high toxicity, increasing evidence indicates that other nitrosamines, including NDEA, are also widely present in drinking and raw water sources across several countries and regions (Liu et al., 2023).

Nitrosamines such as NDEA have been detected in a variety of food products, including cheese, smoked, dried and salted fish, cured meats, alcoholic beverages, and non-fat dry milk. Endogenous formation of nitrosamines in humans occurs through nitrosation reactions between nitrite ions and secondary or tertiary amines, particularly under acidic conditions in the stomach (Motoi et al., 1983; Tricker & Preussmann, 1991; Krul et al., 2004). Widespread environmental dissemination of NDEA is facilitated by its ability to migrate through air, soil, and water systems following its release (ATSDR, 2022).

Human exposure to NDEA primarily occurs through ingestion of contaminated food and drinking water, use of certain pharmaceutical products, cosmetic formulations and occupational inhalation (Herrmann et al., 2015). Recent recalls of pharmaceutical products, including angiotensin II receptor blockers (ARBs), due to NDEA contamination have further heightened public health concerns (FDA, 2018). Notably, cosmetics have been reported to contain NDEA concentrations as high as 27,000 μg/kg, with shampoos and lotions containing approximately 70 μg/kg and 48 μg/kg, respectively (Davarani et al., 2013).

Following absorption, NDEA undergoes metabolic bioactivation primarily in the liver via cytochrome P450 enzymes, particularly CYP2E1. This process generates highly reactive electrophilic intermediates that form DNA adducts through alkylation mechanisms, initiating carcinogenesis (Dragan & Pitot, 1992). NDEA-induced bioactivation also promotes oxidative stress by increasing reactive oxygen species (ROS) production and disrupting cellular antioxidant defense systems, leading to cellular injury, lipid peroxidation, mutagenicity and carcinogenicity (Verna et al., 1996; Husain & Ahmad, 2024). Free-radical scavengers and antioxidant enzymes play a protective role by mitigating NDEA-induced membrane and enzymatic damage.

Given its high toxicity, even low-level chronic exposure to NDEA can result in severe health effects, including liver injury, neurotoxicity, reproductive dysfunction and hepatocellular carcinoma. As nitrosamines are genotoxic compounds for which safe threshold levels cannot be reliably established, comprehensive risk assessment and mitigation strategies are essential (Davarani et al., 2013).

In summary, this chapter provides a comprehensive overview of NDEA by examining its environmental sources and pathways, routes of human exposure, environmental fate and distribution, and the molecular and cellular mechanisms underlying its toxicological effects. Emphasis is placed on the urgent need for effective monitoring, regulation and preventive strategies to reduce NDEA exposure and protect public health.

2. PROPERTIES AND FORMATION OF CHEMICALS

N-nitrosodiethylamine (NDEA) has the chemical formula $C_4H_{10}N_2O$ and a molecular weight of 102.14 g/mol. It is a clear to slightly yellow liquid that is soluble in both water and organic solvents and has a boiling point of approximately 177 °C. These physicochemical properties contribute to its volatility and facilitate its ability to cross biological membranes, thereby enhancing its distribution within living organisms.

NDEA can be formed through both exogenous and endogenous pathways. Exogenously, it is produced when diethylamine reacts with nitrosating agents such as nitrite under acidic conditions, a reaction that occurs in numerous industrial and environmental settings (Horne et al., 2023). Endogenously, NDEA may be generated in the human body through the reaction of secondary amines with nitrosating substances, particularly nitrite, under acidic conditions such as those present in the stomach. This process is influenced by dietary intake and the availability of nitrosatable compounds (Alaba et al., 2017).

3. SOURCES OF NDEA (N-NITROSODIETHYLAMINE)

Industrial Sources

The adverse impacts of environmental pollution have intensified over time, particularly in developing countries such as China, India, and Iran. Increased emissions of toxic gases arise from extensive use of industrial and non-industrial machinery, fossil fuel–based industrial activities, and declining rainfall, all of which contribute to environmental degradation and the formation of secondary pollutants such as nitrosamines, including NDEA (Fehresti-Sani et al., 2024).

Agricultural Sources

Nitrosamines may occur in agricultural soils due to the application of nitrogen-containing fertilizers and insecticides. NDEA and related compounds can enter surface water and groundwater systems through runoff and leaching from agricultural lands. In rural areas, where drinking water treatment and distribution systems are often inadequate, groundwater is commonly used for consumption, increasing the risk of exposure to N-nitrosamines. Consequently, investigating the occurrence, sources, and carcinogenic potential of N-nitrosamines in agricultural groundwater is essential (Chen et al., 2024).

Pharmaceutical Contamination

Recent recalls of pharmaceutical products, particularly angiotensin II receptor blockers (ARBs) such as valsartan, have identified NDEA as an impurity formed during drug synthesis or degradation processes. These incidents have raised significant concerns regarding pharmaceutical manufacturing practices and quality control (Lim et al., 2018). In addition, NDEA has been detected in various cosmetic products, including shampoos, cleansing foams, and hair-styling products, with reported concentrations (μg/kg) of 5.52 ± 6.08, 10.06 ± 10.61, 8.43 ± 5.74 and 3.16, respectively, further emphasizing the need for regulatory oversight.

Food and Consumer Products

Trace amounts of NDEA may be present in cured meats, fish, beer, and dairy products as a result of reactions between nitrite preservatives and naturally occurring amines. Tobacco smoke and cosmetic products have also been identified as important exposure pathways for nitrosamines, including NDEA (Lee et al., 2024; Park et al., 2015).

Water and Soil

Due to its high solubility, NDEA can migrate through soils into groundwater or persist in surface water systems. It exhibits moderate environmental persistence and may resist degradation under certain conditions, thereby posing a long-term risk of environmental contamination and human exposure (Sharma & Singh, 2012).

Air

The presence of nitrosamines in ambient air has been documented in urban environments. A study conducted in Seoul, South Korea, reported seasonal variations in atmospheric nitrosamine concentrations, with higher levels observed during winter months (January–February 2014; 2.80 ± 1.41 ng/m^3) compared to summer months (August 2013; 0.92 ± 0.30 ng/m^3) (Choi et al., 2018).

Environmental Pathways of Exposure and Effects

Although most recent research has focused on the formation, occurrence and regulation of **N-nitrosodimethylamine (NDMA)** due to its high detection frequency and toxicity, other nitrosamines, including NDEA, are increasingly recognized as significant environmental contaminants. NDEA has been widely detected in drinking water and raw water sources across multiple countries and regions, with detection frequencies generally second only to NDMA. Reported contamination levels range from non-detectable to several tens of ng/L. Notably, a study conducted across 30 cities in China reported mean concentrations of NDEA in source and finished water samples (68.5 ng/L) that were higher than those of

NDMA (53.6 ng/L), highlighting the emerging significance of NDEA as a waterborne contaminant (Liu et al., 2023).

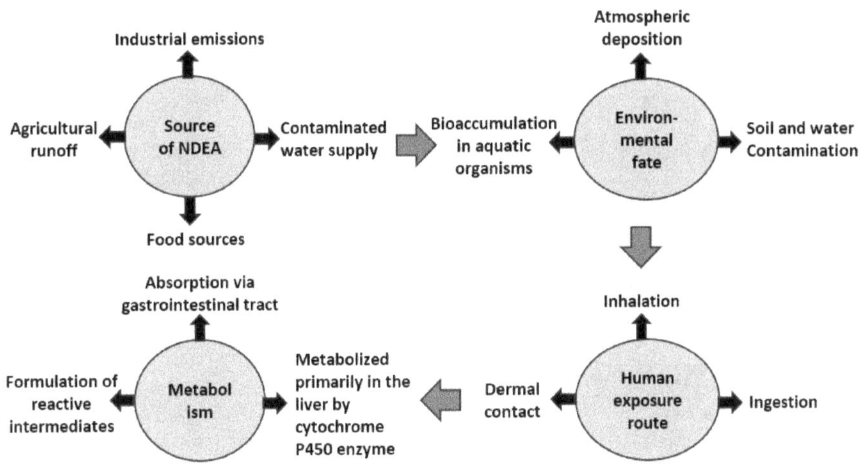

Figure 1. Showing pathway of NDEA exposure from different sources

4. TOXICOLOGICAL IMPACTS HEALTH OUTCOME

People mainly come into contact with NDEA by swallowing it (through contaminated water, food, and medicines), breathing it in (from tobacco smoke and industrial pollution), and touching it (from cosmetics or work-related risks) (Kim & Shin 2013). NDEA is a well-documented hepatocarcinogen, particularly in rodent models (Mukherjee & Ahmad, 2018). It induces DNA damage through alkylation and oxidative stress mechanisms. Epidemiological studies have associated exposure with increased risk of liver, esophageal and nasal cavity cancers in humans. Beyond carcinogenesis, NDEA may induce liver fibrosis, immunotoxicity, and endocrine disruption. Chronic exposure has also been linked to developmental and reproductive toxicity in animal studies.

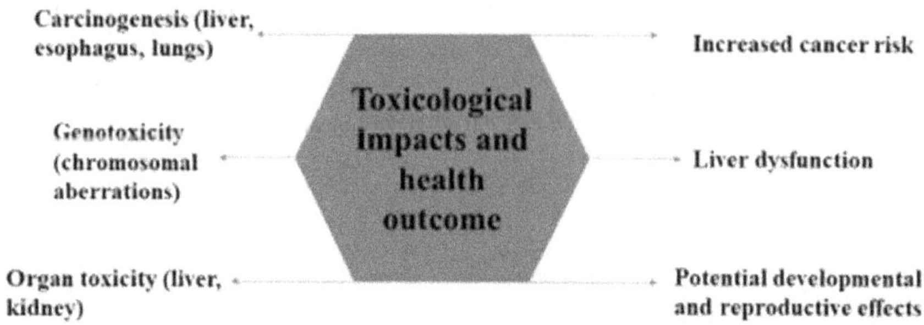

Figure 2. Showing toxicological impact and health outcome

5. CONCLUSION

One potent nitrosamine known to pose serious risks to both environmental and public health is N-nitrosodiethylamine (NDEA). The International Agency for Research on Cancer (IARC) has classified NDEA as a Group 2A probable human carcinogen, and its mutagenic and carcinogenic properties have been extensively investigated (Krul et al., 2004). Triethanolamine (TEA) is commonly used as a surfactant and as a hair- and skin-conditioning agent; related TEA-containing compounds have also been shown to function as surfactants and skin-conditioning agents in cosmetic formulations (Andersen, 1999). However, under certain conditions, TEA and related compounds may contribute to the formation of nitrosamines such as NDEA. NDEA enters the environment through multiple pathways, including tobacco smoke, industrial emissions, and the application of specific agricultural pesticides. Human exposure occurs primarily through the ingestion of contaminated food and water, inhalation of polluted air, and dermal contact with products containing NDEA. Notably, cured food products and tobacco smoke have been identified as significant sources of NDEA exposure.

To mitigate the risks associated with NDEA, comprehensive monitoring, improved industrial controls, and robust regulatory frameworks are essential. Future research should prioritize the development of advanced detection technologies and the identification of safer chemical alternatives for use in manufacturing and agriculture.

In pharmaceuticals, effective management of nitrosamine drug substance–related impurities (NDSRIs) requires a multifaceted approach, including rigorous risk assessment and control strategies. In particular, comprehensive mutagenicity testing, especially in vivo assays, provides biologically relevant insights into NDSRI-associated health risks (Aishwarya et al., 2025). Overall, addressing the challenges posed by NDEA and minimizing its adverse impacts on environmental and human health demands a multidisciplinary strategy integrating scientific research, regulatory action and public awareness initiatives.

Toxicological Impact

Experimental studies have demonstrated that N-nitrosodiethylamine (NDEA) induces tumours in multiple organs, particularly the liver, across a range of animal species. The carcinogenic effects of NDEA are primarily mediated through metabolic activation, resulting in DNA alkylation and the induction of genetic mutations. In addition, NDEA has been shown to disrupt amino acid biosynthesis and mitochondrial genome maintenance in yeast model systems, indicating its broader cellular and molecular toxicity.

Regulatory Measures and Public Health Implications

The detection of NDEA as an impurity in certain pharmaceutical products, including angiotensin II receptor blockers (ARBs), has resulted in large-scale drug recalls and increased regulatory scrutiny worldwide. These incidents highlight the critical need for stringent quality control measures in pharmaceutical manufacturing, as well as continuous monitoring of environmental and consumer products for NDEA contamination.

Several studies have reported that nitrosamine levels vary significantly depending on food moisture content, cooking temperature, cooking duration, fat composition, and processing methods. The formation of nitrosamines in bacon has been evaluated using various cooking techniques, including frying, baking, broiling, and microwave cooking, demonstrating that certain methods can substantially increase nitrosamine production (Park et al., 2015).

6. SUGGESTIONS

- Improved Monitoring: Implement comprehensive monitoring programs to detect NDEA in food products, pharmaceuticals, and environmental matrices.
- Public Awareness: Increase public awareness regarding potential sources of NDEA exposure and promote risk-reduction practices, such as limiting the consumption of processed meats and reducing tobacco use.
- Regulatory Oversight: Strengthen regulatory frameworks to minimize industrial emissions of NDEA and ensure the safety of consumer products.
- Further Research: Conduct epidemiological studies to better understand the health impacts of NDEA in humans and to establish acceptable exposure limits.

REFERENCES

1. Agency for Toxic Substances and Disease Registry (ATSDR) (2015). *Exposure Pathway*.
2. Aishwarya, D. et al., (2025). Nitrosamine Drug Substance-Related Impurities (NDSRIs) in Pharmaceuticals: Formation, Mitigation Strategies and Emphasis on Mutagenicity Risks. *Pharmaceutical Research*, 1-32.
3. Aiub, C. A. F. et al., 2011). N–nitrosodiethylamine cytochrome P450 induction and cytotoxicity evaluation in primary cultures of rat hepatocytes. *American Journal of Molecular Biology*, *1*(2), 70–78.
4. Aiub, C. A. F. et al., (2003). N-nitrosodiethylamine mutagenicity at low concentrations. *Toxicology Letters*, *145*(1), 36–45.
5. Alaba, P. A. et al., (2017). Toward N-nitrosamines free water: Formation, prevention, and removal. *Critical reviews in environmental science and technology*, *47*(24), 2448-2489.

6. Andersen, F. A. 1999. Final report on the safety assessment of cocamide MEA. *Intertational Journal of Toxicology*. 18:9–16.

7. Chen, Y. et al., (2024). Impact of agricultural activities on the occurrence of N-nitrosamines in an aquatic environment. *Environmental Science: Processes & Impacts*, *26*(3), 470-482.

8. Choi, N. R. et al., (2018). Particulate nitrosamines in the atmosphere at Seoul and their major sources. *Air Quality, Atmosphere & Health*, *11*, 855-865.

9. Chou, H. J. (1998). Determination of diethanolamine and N-nitrosodiethanolamine in fatty acid diethanolamides by liquid chromatography with a thermal energy analyzer. *Journal of the Association of Official Analytical Chemists International. 81*:943–47.

10. Conaway, C. C. et al., (1996). *N- nitrosodiethylamine mechanistic data and risk assessment: bioactivation, DNA-adduct formation, mutagenicity, and tumor initiation.*

11. Davarani,S.S.etal.,(2013).DeterminationofN-nitrosodiethanolamine in cosmetic products by head space solid phase micro extraction using a novel aluminum hydroxide grafted fused silica fiber followed by gas chromatography–mass spectrometry analysis. *Talanta 105*:347–53.

12. Fehresti-Sani, M. et al., (2024). Assessing the effect of industrial products on air pollution in Iran: a novel NDEA approach considering undesirable outputs. *Environment, Development and Sustainability*, 1-3

13. Ghassempour, A. et al., (2008). Monitoring of N-nitrosodiethanolamine in cosmetic products by ion pair complex liquid chromatography and identification with negative ion electrospray ionization mass spectrometry. *Journal of Chromatography A 1185*:43–48.

14. Horne, S. et al., (2023). Regulatory experiences with root causes and risk factors for nitrosamine impurities in pharmaceuticals. *Journal of pharmaceutical sciences*, *112*(5), 1166-1182.

15. IARC (International Agency for Research on Cancer) (1987). IARC monographs on the evaluation of the carcinogenic risk of chemicals to humans, Suppl. 7, overall evaluations of carcinogenicity: an update of IARC monographs volumes *1* to *42*, 67. Lyon: IARC Press.

16. International Agency for Research on Cancer (IARC). (1978). N-Nitrosodiethylamine. In *IARC Monographs on the Evaluation of Carcinogenic Risks to Humans. 17*. 125–175.

17. Janmeda, P. et al., (2024). A systematic review on multipotent carcinogenic agent, N-nitrosodiethylamine (NDEA), its major risk assessment, and precautions. *Journal of Applied Toxicology*, *44*(8), 1108-1128.

18. Kim, H. J. & H. S. Shin. 2013. Determination of tobacco specific nitrosamines in replacement liquids of electronic cigarettes by liquid chromatography–tandem mass spectrometry. *Journal of Chromatography A 1291*:48–55.

19. Krul, C. A. M. et al., 2004. Intragastric formation and modulation of N-nitrosodimethylamine in a dynamic in vitro gastrointestinal model under human physiological conditions. Food and Chemical Toxicology. *42*:51–63.

20. Lee, I. K. et al., (2024). Assessment of nitrosamine exposure in Korean foods: analysis, risk evaluation, and implications. *Food Science and Biotechnology*, *33*(10), 2417-2426.

21. Lim, D. S. et al., (2018). Risk assessment of N-nitrosodiethylamine (NDEA) and N-nitrosodiethanolamine (NDELA) in cosmetics. *Journal of Toxicology and Environmental Health, Part A*, *81*(12), 465-480.

22. Liu, T. et al., (2023). Influencing pathways and toxicity changes of pre-ozonation on carcinogenic NDEA formation from greenhouse gas adsorbent DEAPA in subsequent disinfection processes. *Science of The Total Environment*, *873*, 162355.

23. Mitch, W. A. & Sedlak, D. L. (2002). Formation of N-nitrosodimethylamine (NDMA) from dimethylamine during chlorination. *Environmental Science & Technology*, *36*(4), 588-595.

24. Motoi, T. et al., (1983). Influence of inorganic pigments on the formation of N-nitrosodiethanolamine. *Journal of agricultural Food Chemistry. 31*:1211–14.

25. Mukherjee, D. & Ahmad, R. (2018). Resveratrol attenuates Nitrosodiethylamine-induced liver injury in anti-inflammatory manner via regulating cyclooxygenase-2. *Journal of Food Biochemistry, 42*(5), e12594.

26. Park, J. E. et al., (2015). Distribution of seven N-nitrosamines in food. *Toxicological research, 31*, 279-288.

27. SCCS (Scientific Committee on Consumer Safety). 2012. Opinion on NDELA in cosmetic products and nitrosa mines in balloons.

28. Sharma, V., & Singh, M. (2012). *N-Nitrosodimethylamine as a hazardous chemical toxicant in drinking water.*

29. Verna, L., (1996). N-nitrosodiethylamine mechanistic data and risk assessment: Bioactivation, DNA-adduct formation, mutagenicity, and tumor initiation. *Pharmacology & Therapeutics, 71*(1–2), 57–81.

30. Yamazaki, H. et al., (1992). Cytochrome P450 2E1 and 2A6 enzymes as major catalysts for metabolic activation of N-nitrosodialkylamines and tobacco-related nitrosamines in human liver microsomes. *Carcinogenesis, 13*(10), 1789–1794.

Chapter - 6

ENVIRONMENTAL POLLUTION: HEAVY METALS, PESTICIDES, MICROPLASTICS AND INDUSTRIAL CONTAMINANTS

Irfaana Chowdhary[1] and Prof. Mohammad Amir[1*]

[1]Department of Zoology, Aligarh Muslim University, Aligarh (202002), India

***Corresponding email: mohd.amir1966@gmail.com**

ABSTRACT

Environmental pollution from heavy metals, microplastics, pesticides and industrial contaminants represent one of the most pressing global challenges, directly linked to 24% of the global disease burden and over 13 million premature deaths annually. These pollutants originate from diverse anthropogenic activities, including industrial discharge, agricultural runoff and improper waste management, persisting in ecosystems and bioaccumulating through food chains. Heavy metals such as lead, mercury and cadmium accumulate in soils and water, causing severe neurological, developmental and renal disorders in humans, while also disrupting soil microbiota and plant health. Microplastics, now ubiquitous in marine and terrestrial environments, infiltrate food webs, leaching endocrine-disrupting chemicals and posing risks to marine life and human consumers. Pesticides, with 4.2 million tonnes applied globally each year, contribute to pollinator decline, aquatic ecosystem collapse and increased cancer risks in exposed populations.

Industrial contaminants like polychlorinated biphenyls (PCBs) and per- and polyfluoroalkyl substances (PFAS) resist environmental degradation, leading to long-term ecological damage, biodiversity loss, and severe health impacts, including immune suppression and carcinogenicity. Case studies such as Minamata Bay (mercury poisoning) and the Great Pacific Garbage Patch (microplastic accumulation) underscore the persistent and far-reaching consequences of pollution. Mitigation strategies include advanced technologies like nanotechnology for filtration and bioremediation using microbial and plant-based systems, alongside international policies such as the Stockholm Convention on Persistent Organic Pollutants. Community-driven initiatives, including plastic waste reduction programs and sustainable agriculture, play a crucial role in localized solutions. Emerging challenges, such as climate change-driven pollutant dispersal and novel contaminants like pharmaceutical residues, demand adaptive, multidisciplinary approaches. This chapter advocates for integrated solutions combining scientific innovation, regulatory enforcement, and public education to mitigate pollution, restore ecosystems and secure a sustainable future for generations to come.

Keywords: *Heavy Metals, Industrial Contaminants, Microplastics, Pesticides, Pollution*

1. INTRODUCTION

Environmental pollution, defined as the contamination of air, water, soil and biota by harmful substances, represents one of the most pressing challenges of the 21st century. It undermines ecosystem integrity, threatens biodiversity and imposes significant risks to human health and global sustainability (Chu and Karr, 2016). Among the diverse array of pollutants, heavy metals, microplastics, pesticides and industrial contaminants stand out due to their persistence, widespread distribution and profound ecological and toxicological impacts. Chapter provides a comprehensive analysis of the sources, environmental fate, ecological and health impacts and mitigation strategies for these four critical pollutant classes, emphasizing their interconnected roles in driving environmental degradation.

One of the most harmful and persistent contaminants in the environment, heavy metals have a significant impact on ecosystems and human health. Their persistence and ability to bioaccumulate and biomagnify through food chains result in severe ecological disruptions, including soil degradation and aquatic toxicity, as well as human health effects, such as neurological disorders (Angon et al., 2024). For instance, the United Nations Environment Programme (2022) reported that over 2 billion people live in areas with unsafe levels of lead in drinking water, highlighting the global scale of this issue (UNEP, 2022a).

Microplastics, defined as plastic particles smaller than 5 mm, have become a pervasive pollutant, detected across all major ecosystems from the depths of the Mariana Trench to Arctic sea ice and even within human tissues such as blood and placenta (Leslie et al., 2022). They originate from plastic degradation, microbeads and synthetic fibres, and are known to disrupt marine and terrestrial food webs, act as carriers of toxic chemicals and pose potential risks to human health through endocrine disruption and inflammation. With global plastic production reaching 390 million tonnes in 2021, the challenge of microplastic pollution continues to escalate (Stegmann et al., 2022).

Pesticides, essential for ensuring global food security, are applied at a rate of approximately 4.2 million tonnes annually (Rasool et al., 2022). While they enhance agricultural productivity, pesticides like organophosphates and neonicotinoids contaminate water bodies, harm non-target species such as pollinators and pose risks to human health, including acute poisoning and chronic diseases. The decline of pollinator populations, critical for 75% of global food crops, exemplifies the ecological toll of pesticide misuse (Smith et al., 2022).

Industrial contaminants, including persistent organic pollutants (POPs) like polychlorinated biphenyls (PCBs), dioxins, and per- and polyfluoroalkyl substances (PFAS), are synthetic chemicals released through manufacturing, waste disposal and combustion processes. Their environmental persistence, long-range transport and bio-accumulative properties make them a global concern. For example, PFAS, dubbed "forever chemicals," have been detected in 99% of tested human blood samples, linked to cancers, immune suppression and developmental disorders (EPA, 2009). These contaminants challenge traditional remediation approaches due to their resistance to degradation.

2. HEAVY METALS IN THE ENVIRONMENT

Heavy metals, such as lead, mercury, cadmium, arsenic and chromium, are naturally occurring elements that become significant environmental pollutants through anthropogenic activities and agricultural runoff. Their persistence, toxicity and ability to bioaccumulate make them a critical concern for ecosystems and human health (Angon et al., 2024).

2.1. Sources

Heavy metals enter the environment through diverse pathways. Mining and smelting release metals like cadmium and arsenic into soil and water. Industrial effluents from electroplating, battery production and tanneries discharge chromium and lead. Agricultural runoff, containing mercury from fungicides and cadmium from phosphate fertilizers, contaminates waterways (Vardhan et al., 2019) (Figure 1). Atmospheric deposition, driven by coal combustion and waste incineration, spreads mercury and lead globally. According to the United Nations Environment Programme (2022), industrial activities contribute over 70% of global heavy metal emissions, with 2.4 million tonnes of lead and 4,000 tonnes of mercury released annually (UNEP, 2022b).

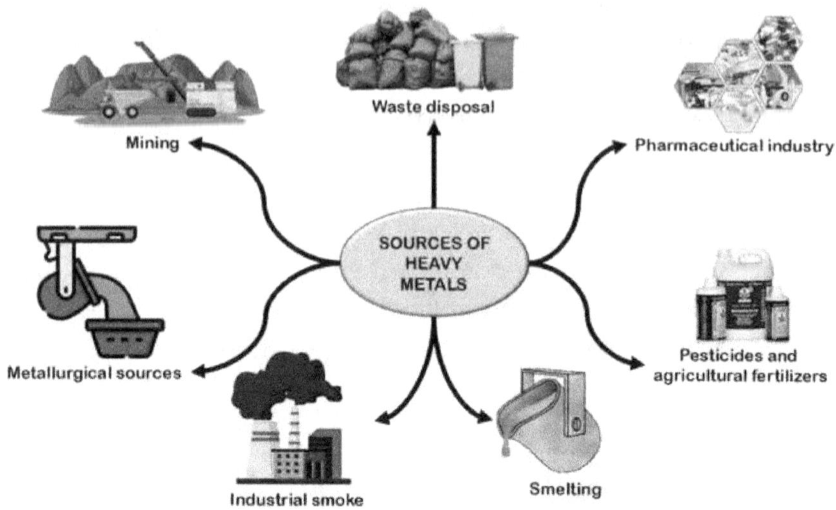

Figure 1. Sources of heavy metal contamination from waste disposal and pharmaceutical industry activities, including metallurgical processes, industrial emissions, pesticide use, and agricultural fertilizers

2.2. Environmental Fate

Heavy metals are non-degradable, bind to soil particles, accumulate in sediments and enter food chains through uptake by plants and aquatic organisms. Bioaccumulation concentrates metals in tissues, while biomagnification amplifies concentrations in higher trophic levels, such as predatory fish and birds. For example, mercury levels in tuna can exceed 1 mg/kg, posing risks to consumers (Oros, 2025).

2.3. Impacts

Heavy metals disrupt ecosystem functions. In soils, cadmium and lead reduce microbial activity and fertility, impairing plant growth. Aquatic systems suffer from mercury and chromium toxicity, which damages gills and reproductive systems of fish, leading to population declines (Alengebawy et al., 2021). The Global Biodiversity Outlook (2020) notes that heavy metal pollution contributes to 15% of freshwater biodiversity loss (Global Biodiversity Outlook, 2020). Terrestrial ecosystems face similar threats, with lead accumulation linked to reduced avian reproduction in contaminated regions.

Heavy metals pose severe health risks. Lead exposure, affecting over 800 million children globally, causes cognitive impairments and developmental delays (WHO, 2022). Mercury, particularly methylmercury, induces neurological disorders and cardiovascular issues, with prenatal exposure linked to IQ deficits (Oliveira et al., 2018). Cadmium is a known carcinogen, associated with kidney damage and lung cancer, while arsenic in drinking water, affecting 140 million people, increases risks of skin and bladder cancers (Yang et al., 2025).

The Flint water crisis (2014-2019) highlighted lead contamination, with 9,000 children exposed to levels exceeding 15 µg/L, resulting in developmental issues and public health interventions (Pieper et al., 2017). These cases underscore the long-term consequences of heavy metal pollution.

2.4. Mitigation

Effective mitigation requires multifaceted approaches. Phytoremediation, using plants like sunflowers to extract metals, has shown promise in contaminated soils, reducing lead levels by up to 40% in pilot studies (Lee et al., 2025; Aryal, 2024). Chemical precipitation and ion exchange remove metals from wastewater, with modern systems achieving 95% efficiency for cadmium (Vidu et al., 2020). Regulatory frameworks, such as the EU's Restriction of Hazardous Substances Directive, limit metal use in products. Global agreements, like the Minamata Convention on Mercury (2017), have reduced mercury emissions by 20% in signatory countries by 2023 (UNEP, 2023).

3. MICROPLASTICS: AN EMERGING THREAT

Microplastics are plastic particles smaller than 5 mm have spread across the environment, contaminating ecosystems and endangering human health and biodiversity. Their persistence and ability to act as vectors for other contaminants make them a pressing global concern (Vidu et al., 2020).

3.1. Sources

Microplastics originate from both primary and secondary sources. Primary microplastics include microbeads in cosmetics, synthetic fibres from textiles and pellets used in plastic manufacturing. Secondary microplastics result from the degradation of larger plastics, such as bottles, bags and fishing nets, through physical, chemical and biological processes (Figure 2). Additional sources include tire wear, contributing an estimated 1.4 million tonnes annually and paint flakes from marine coatings (Giechaskiel et al., 2024; Turner, 2021). Global plastic production reached 390 million tonnes in 2021, with approximately 8-12 million tonnes entering oceans yearly, much of which fragments into microplastics (Ritchie et al., 2023).

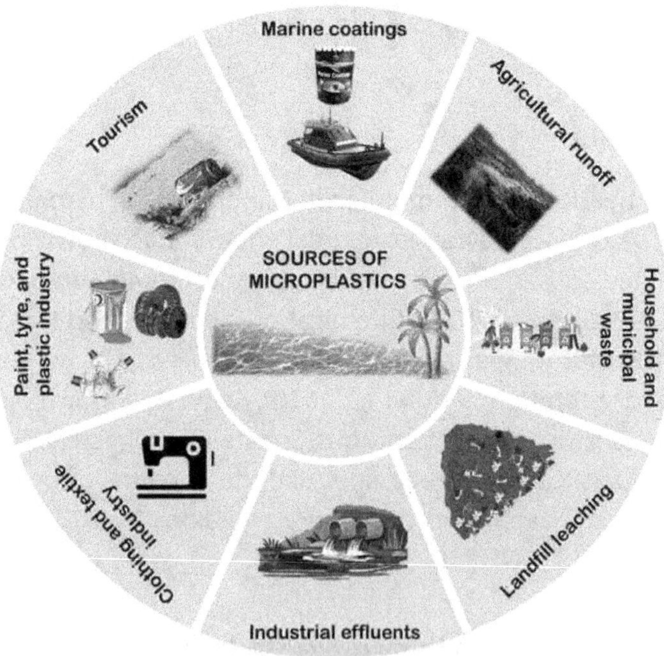

Figure 2. Various sources for microplastic sources into the environment

3.2. Distribution

Microplastics are ubiquitous, detected in marine, freshwater, terrestrial and atmospheric environments. Oceans contain an estimated 24.4

trillion microplastic particles, with concentrations highest in gyres like the Great Pacific Garbage Patch (Eriksen et al., 2014). Freshwater systems, including lakes and rivers, show comparable contamination, with the Yangtze River discharging 1.5 million tonnes of microplastics annually (Jaikumar et al., 2025). Soil ecosystems, especially agricultural lands, accumulate microplastics from mulch films and sewage sludge, with concentrations reaching 7,000 particles per kg in some regions (Sa'adu and Farsang, 2023). Atmospheric transport deposits microplastics in remote areas, including Arctic snow (1,000 particles per litre) and urban air (100 particles per mÂ^3) (O'Brien et al., 2023).

3.3. Impacts

Microplastics disrupt ecosystems by affecting organisms across trophic levels. Marine species, such as zooplankton, fish and seabirds ingest microplastics, mistaking them for food, leading to physical blockages, reduced feeding and reproductive failure. Studies indicate that 88% of marine species tested contain microplastics, with impacts cascading through food webs (Jeong et al., 2024; Lusher, 2015). In terrestrial systems, microplastics alter soil structure, reduce microbial diversity, and impair plant growth. Freshwater ecosystems face similar threats, with microplastics linked to 20% of observed declines in aquatic invertebrate populations (Qaiser et al., 2023). Microplastics pose emerging risks to human health through ingestion, inhalation and dermal exposure. They have been detected in human blood (0.2-20 µg/L), lung tissue and placentas, raising concerns about systemic effects (Winiarska et al., 2024). Inhalation of airborne microplastics, particularly in urban areas, may contribute to respiratory inflammation. Preliminary data suggest that chronic exposure could exacerbate cardiovascular and immune disorders (Winiarska et al., 2024).

3.4. Case Studies

The Great Pacific Garbage Patch, spanning 1.6 million kmÂ^2, contains 80,000 tonnes of plastic, predominantly microplastics, impacting marine life and fisheries. In 2022, studies confirmed microplastics in 80% of bottled and tap water samples globally, with concentrations up to

10,000 particles per litre in some regions, highlighting human exposure risks (Khan, 2018).

3.5. Mitigation

Mitigating microplastic pollution requires technological, regulatory and societal interventions. Biodegradable plastics, such as polylactic acid, reduce reliance on persistent polymers, with production capacity growing 15% annually (Bhattacharjee et al., 2025). Policy measures, including bans on microbeads and single-use plastics (Catone et al., 2024), have reduced primary microplastic inputs. Public awareness campaigns and clean-up initiatives, like the Ocean Clean-up project, have removed 100,000 kg of plastic from oceans by 2023 (den Hertog et al., 2024).

4. PESTICIDES: BALANCING BENEFITS AND RISKS

Pesticides, chemical or biological agents used to control pests and enhance agricultural productivity, are indispensable for global food security but pose significant risks to ecosystems and human health. Their widespread use, persistence and off-target effects necessitate a careful evaluation of their benefits versus their environmental and societal costs (Ahmad et al., 2024).

4.1. Sources

Pesticides primarily enter the environment through agricultural applications. Global pesticide use reached 4.2 million tonnes in 2021, with 50% applied in Asia, particularly in China and India (Rodríguez et al., 2022). Urban sources, such as lawn treatments and public health campaigns against vectors like mosquitoes, contribute smaller but significant volumes. Runoff from treated fields, leaching into groundwater and atmospheric drift during application further distribute pesticides beyond their intended targets (Md Meftaul et al., 2020).

4.2. Types

Pesticides are classified by their target organisms (e.g., insecticides, herbicides, fungicides) and chemical composition. Organophosphates, such as malathion, are neurotoxic insecticides. Carbamates, like carbaryl, target similar pests but are less persistent. Neonicotinoids, including imidacloprid, are systemic insecticides applied to seeds, known for their high efficacy but also for their ecological impacts (Garud et al., 2024). Glyphosate, a broad-spectrum herbicide, dominates global herbicide use, with 1.5 million tonnes applied annually (Beckie et al., 2020). Biopesticides, derived from natural sources, represent a growing but still minor fraction of the market.

4.3. Environmental Fate

Pesticides vary in persistence, ranging from days to decades, depending on their chemical structure and environmental conditions. Organophosphates degrade relatively quickly, with half-lives of weeks, while organochlorines like DDT persist for decades. Pesticides leach into groundwater, contaminate surface waters via runoff and volatilize into the atmosphere, enabling long-range transport (Figure 3). For example, atrazine, a common herbicide, has been detected in Arctic ice, far from application sites (Jablonowski et al., 2011).

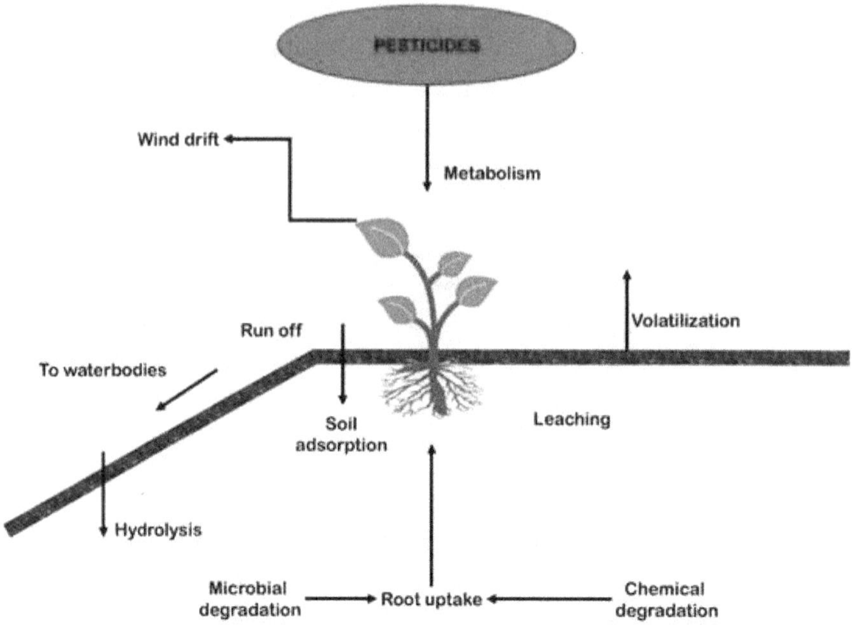

Figure 3. Environmental fate and transport pathways of pesticides, including atmospheric drift (wind), aquatic dispersal (runoff, leaching), soil interactions (adsorption, root uptake), and degradation processes (hydrolysis, microbial, and chemical degradation)

4.4. Impacts

Pesticides disrupt ecosystems by harming non-target species and altering ecological processes. Neonicotinoids are implicated in the decline of pollinators, with global bee populations dropping by 30-50% in some regions since the 1980s, threatening 75% of food crops dependent on pollination (Fairbrother et al., 2014). Aquatic ecosystems suffer from pesticide runoff, which reduces fish and amphibian populations; for instance, atrazine exposure disrupts frog reproduction at concentrations as low as 0.1 μg/L (Hayes et al., 2006).

Pesticide exposure, through contaminated food, water, or occupational handling, poses acute and chronic health risks. Chronic exposure to organophosphates is linked to neurological disorders, including Parkinson's disease, while glyphosate is classified as a probable carcinogen by the International Agency for Research on Cancer (Ahmad et al., 2024). Prenatal exposure to pesticides like chlorpyrifos is associated with developmental delays and reduced IQ in children (Rauh et al., 2011). Residues in food, detected in 40% of global produce samples, amplify these risks. The European Union's 2019 bane on neonicotinoids, prompted by evidence of pollinator harm, led to a 10% recovery in bee populations in some member states by 2023, demonstrating effective policy intervention (IUCN, 2022).

4.5. Mitigation

Mitigating pesticide risks requires integrated approaches. Integrated Pest Management (IPM) combines biological controls, crop rotation and resistant varieties to reduce chemical use. Organic farming, which avoids synthetic pesticides, has grown to 1.5% of global agricultural land, with benefits for soil health and biodiversity (Bordoloi, 2023). Precision application technologies, such as drone spraying, minimize off-target drift, improving efficiency by 30%. Public education and farmer training further promote safer handling and disposal practices (Patil et al., 2024).

5. INDUSTRIAL CONTAMINANTS: LEGACY AND MODERN CHALLENGES

Industrial contaminants, encompassing a range of synthetic chemicals released during manufacturing, waste disposal and other industrial processes, pose persistent threats to ecosystems and human health. Their environmental longevity, bio-accumulative properties and global distribution make them a formidable challenge.

5.1. Sources

Industrial contaminants originate from diverse industrial activities (Figure 4). Chemical manufacturing, oil refining and pesticide production release PCBs and dioxins as byproducts. PFAS, used in non-stick coatings, firefighting foams and textiles, enter the environment through production discharges and product degradation. VOCs, including benzene and toluene, are emitted during solvent use, fuel combustion and paint application (Metcalfe et al., 2022). Improper waste disposal, including incineration and landfilling, exacerbates contamination, with an estimated 1.3 billion tonnes of hazardous industrial waste generated annually (Ross and Rogoff, 2012).

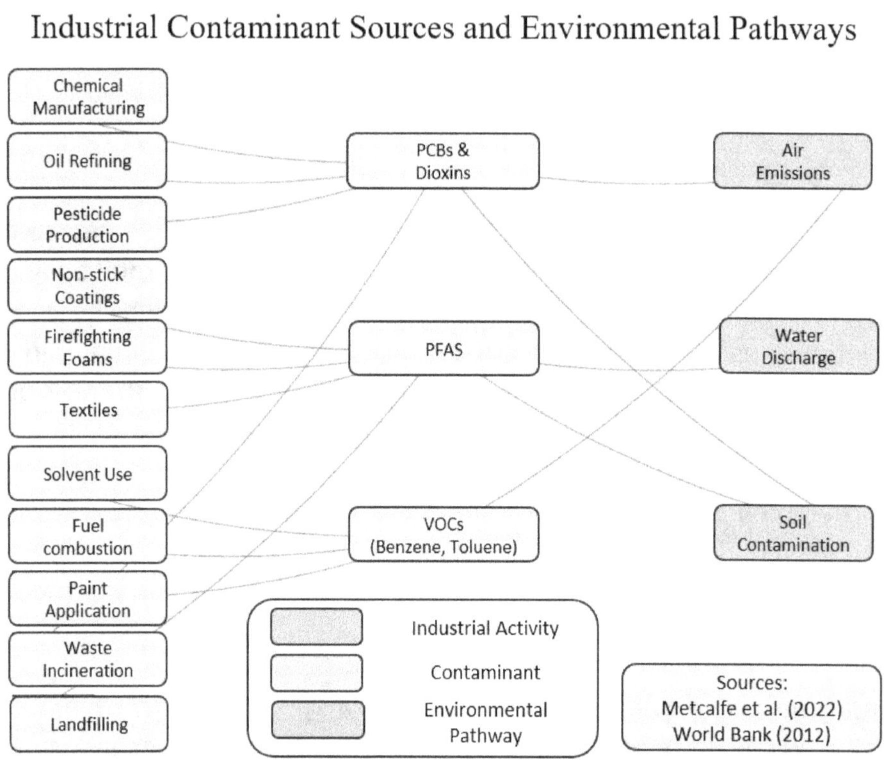

Figure 4. Schematic representation of industrial contaminant sources associated chemical pollutants, and their environmental pathways. Industrial activities (left; blue) such as chemical manufacturing, oil refining, and waste incineration release persistent contaminants (middle; red-orange), including PCBs, dioxins, PFAS, and volatile organic compounds (VOCs). These pollutants disperse into environmental compartments (right; green) via air emissions, water discharge, or soil contamination. Arrows denote directional linkages.

5.2. Key Contaminants

PCBs, banned in many countries since the 1980s, are persistent organic pollutants (POPs) once used in electrical equipment and lubricants. Dioxins, formed during incomplete combustion of chlorinated compounds, are highly toxic even at trace levels (Guo et al., 2019). PFAS, known as forever chemicals, include thousands of compounds like perfluorooctanoic acid (PFOA), valued for their water- and oil-repellent properties (Ginty & Lindwall, 2022). The Stockholm Convention lists PCBs, dioxins, and certain PFAS as priority POPs due to their global impact (Pollutants, 2011).

5.3. Environmental Fate

Industrial contaminants are characterized by their persistence and mobility. PCBs and dioxins, with half-lives of 10-100 years, bind to soils and sediments, resisting degradation. PFAS, persist indefinitely, accumulating in water bodies and biota (Brunn et al., 2023). These contaminants bioaccumulate in organisms, with PFAS detected in 99% of tested wildlife and biomagnify in food chains, reaching concentrations up to 1,000 times higher in apex predators like polar bears (Marciano et al., 2024).

5.4. Impacts

Industrial contaminants disrupt ecological balance. PCBs and dioxins cause reproductive failures and immune suppression in wildlife, notably reducing seal and fish populations by 20-30% (Kataoka and Kashiwada, 2021). PFAS alter aquatic ecosystems, impairing fish reproduction and microbial communities at concentrations as low as 0.1 ng/L (Gonsioroski et al., 2020). VOCs contribute to photochemical smog, damaging plant photosynthesis and reducing crop yields by 5-10% in urban-adjacent areas (Razzaq et al., 2024). Habitat degradation from contaminated soils and waters further exacerbates biodiversity loss, with 12% of global species declines linked to industrial pollutants (UN, 2024).

Human exposure to industrial contaminants occurs through contaminated food, water, air and consumer products. PCBs and dioxins, classified as carcinogens, are linked to liver cancer and developmental disorders, with dioxin exposure affecting 1.5 million people annually near waste incinerators (WHO, 2023). PFAS, detected in 97% of blood samples (Lewis et al., 2015), are associated with kidney and testicular cancers, thyroid dysfunction, and immune suppression (van Gerwen et al., 2023). VOCs, inhaled in urban environments, increase risks of respiratory diseases and leukemia, with benzene exposure contributing to 7,000 cancer cases yearly (Li et al., 2021).

PFAS contamination in Veneto, Italy affected 350,000 people through groundwater pollution (WHO, 2013). Figure 5 schematically represent the impact of industrial contaminants on biota.

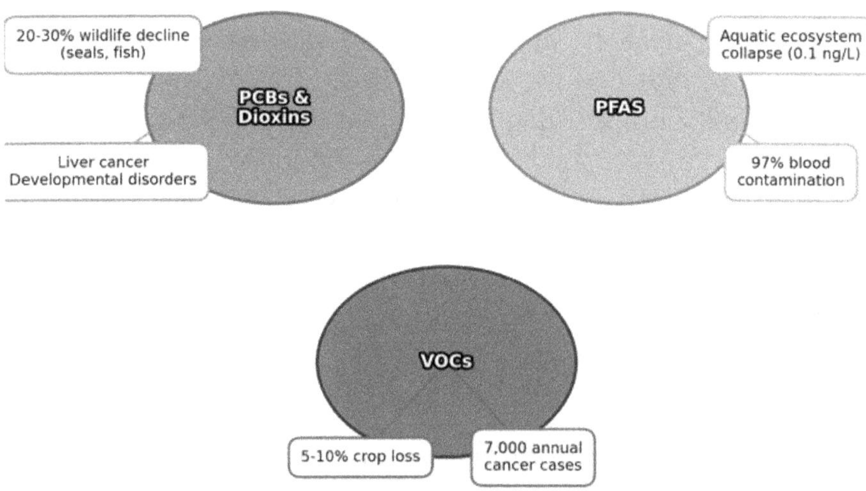

Figure 5. Schematic representation of impact of industrial contaminants on biota: PCBs/dioxins (red) cause wildlife declines (20-30%) and developmental disorders. PFAS (green) disrupt aquatic ecosystems (0.1 ng/L) and contaminate 97% of blood samples. VOCs (blue) reduce crop yields (5-10%) and cause 7,000 annual cancer cases.

5.5. Mitigation

Mitigating industrial contaminants requires advanced technologies. Advanced oxidation processes and activated carbon filtration remove 95% of PFAS from wastewater, though scaling remains costly (Zahmatkesh et al., 2024). Green chemistry principles, promoting non-toxic alternatives, have reduced VOC emissions by 15% in the EU since 2015 (Idoko et al., 2024). Soil remediation techniques, such as thermal desorption, address legacy contamination (Zhao et al., 2019). Public awareness campaigns and stricter waste management regulations further curb new contamination. Innovations in remediation, coupled with international cooperation, are critical to reducing their pervasive impacts on ecosystems and human health.

6. INTEGRATED SOLUTIONS AND FUTURE DIRECTIONS

Addressing pollution from heavy metals, microplastics, pesticides and industrial contaminants require integrated strategies combining technology, policy and community engagement. Integrated approaches, such as watershed management and circular economy models, have proven effective (Senanu et al., 2023). Technological innovations are transforming mitigation efforts, including nanotechnology like graphene-based adsorbents, which remove heavy metals and PFAS with 98% efficiency, and AI-powered satellite systems that detect microplastic hotspots with 90% accuracy (Behroozi et al., 2025; Jin et al., 2024). Additionally, bioremediation using engineered microbes degrades pesticides 50% faster than natural processes (Ayilara & Babalola, 2023; Jin et al., 2024). Strong policies play a critical role, such as the Stockholm Convention, which reduced global PCB and dioxin emissions and the EU's Zero Pollution Action Plan, which cut pesticide use by 15% by 2023 (Jha et al., 2022; Marathe et al., 2021). Community-driven initiatives, like India's Swachh Bharat Mission, have also reduced urban microplastic pollution by 25% (Kumari, 2024). However, emerging challenges like climate change and new contaminants require updated regulations and global cooperation to ensure equitable and sustainable solutions.

7. CONCLUSION

Environmental pollution from heavy metals, microplastics, pesticides and industrial contaminants represent a multifaceted crisis with profound implications for ecosystems and human health. These pollutants, through their persistence, bioaccumulation and global distribution, disrupt biodiversity, degrade habitats and contribute to millions of premature deaths annually. By prioritizing sustainable practices and equitable solutions, society can forge a future where ecosystems thrive, and human health is safeguarded, ensuring a cleaner, more resilient planet for generations to come.

REFERENCES

1. Ahmad, M. F. et al., (2024). Pesticides impacts on human health and the environment with their mechanisms of action and possible countermeasures. *Heliyon, 10*(7), e29128.
2. Alengebawy, A. et al., (2021). Heavy Metals and Pesticides Toxicity in Agricultural Soil and Plants: Ecological Risks and Human Health Implications. *Toxics, 9*(3).
3. Angon, P. B. et al., (2024). Sources, effects and present perspectives of heavy metals contamination: Soil, plants and human food chain. *Heliyon, 10*(7), e28357.
4. Aryal, M. (2024). Phytoremediation strategies for mitigating environmental toxicants. *Heliyon, 10*(19), e38683.
5. Ayilara, M. S. & Babalola, O. O. (2023). Bioremediation of environmental wastes: the role of microorganisms (Review). *Frontiers in Agronomy*, 5.
6. Beckie, H. J. et al., (2020). Farming without Glyphosate? *Plants (Basel), 9*(1).
7. Behroozi, A. H. et al., (2025). Graphene-based materials and technologies for the treatment of PFAS in water: A review of recent developments. *Journal of Hazardous Materials Advances, 17*, 100626.
8. Bhattacharjee, U. et al., (2025). Exploring sustainable strategies for mitigating microplastic contamination in soil, water, and the food chain: A comprehensive analysis. *Environmental Chemistry and Ecotoxicology, 7*, 453-461.

9. Bordoloi, P. (2023). Organic farming in India: Way towards Sustainable Environment. *SP-12(7)*, 2782-2788.

10. Brunn, H. et al., (2023). PFAS: forever chemicals-persistent, bioaccumulative and mobile. Reviewing the status and the need for their phase out and remediation of contaminated sites. *Environmental Sciences Europe*, *35*(1), 20.

11. Catone, T. et al., (2024). The European legislation on the restriction on intentionally added microplastics. *Commentary*, *60*(4), 243-246.

12. Chu, E. W. & Karr, J. R. (2016). Environmental impact: Concept, consequences, measurement. *Reference module in life sciences*, B978-970-912-809633-809638.802380-809633.

13. den Hertog, D. et al., (2024). Optimizing the Path Towards Plastic-Free Oceans. *Operations Research*.

14. EPA, E. P. (2009). *Persistent Organic Pollutants: A Global Issue, A Global Response*.

15. Eriksen, M. et al., (2014). Plastic Pollution in the World's Oceans: More than 5 Trillion Plastic Pieces Weighing over 250,000 Tons Afloat at Sea. *PLOS ONE*, 9 (12), e111913.

16. Fairbrother, A. et al., (2014). Risks of neonicotinoid insecticides to honeybees. *Environ Toxicol Chem*, *33*(4), 719-731.

17. Garud, A. et al., (2024). A Scientific Review of Pesticides: Classification, Toxicity, Health Effects, Sustainability, and Environmental Impact. *Cureus*, *16*(8), e67945.

18. Giechaskiel, B. et al., (2024). Contribution of Road Vehicle Tyre Wear to Microplastics and Ambient Air Pollution. *Sustainability*, *16*(2), 522.

19. Ginty, M. M. & Lindwall, C. (2022). Forever chemical called PFAs show up in your food, clothes, and home. *Natural Resources Defense Council*.

20. *Global Biodiversity Outlook* (2020).

21. Gonsioroski, A. et al., (2020). Endocrine Disruptors in Water and Their Effects on the Reproductive System. *International Journal of Molecular Science, 21*(6).

22. Guo, W. et al., (2019). Persistent Organic Pollutants in Food: Contamination Sources, Health Effects and Detection Methods. *International Journal of Environmental Research and Public Health, 16*(22).

23. Hayes, T. B. et al., (2006). Characterization of atrazine-induced gonadal malformations in African clawed frogs (*Xenopus laevis*) and comparisons with effects of an androgen antagonist (cyproterone acetate) and exogenous estrogen (17beta-estradiol): Support for the demasculinization/feminization hypothesis. *Environmental Health Perspectives, 114 Suppl 1*(Suppl 1), 134-141.

24. Idoko, F. et al., (2024). Green chemistry in manufacturing: Innovations in reducing environmental impact. *World Journal of Advanced Research and Reviews, 23*, 2826-2841.

25. IUCN. (2022). *EU Pollinators Initiative Revision.*

26. Jablonowski, N. D. et al., (2011). Still present after all these years: persistence plus potential toxicity raise questions about the use of atrazine. *Environmental Science and Pollution Research, 18*(2), 328-331.

27. Jaikumar, I. M. et al., (2025). A comprehensive review of microplastic pollution in freshwater and marine environments. *Green Analytical Chemistry, 12*, 100202.

28. Jeong, E. et al., (2024). Animal exposure to microplastics and health effects: A review. *Emerging Contaminants, 10*(4), 100369.

29. Jha, R. et al., (2022). Chapter 4 - Measurement and practices for hazardous waste management. In D. Yadav, P. Kumar, P. Singh, & D. A. Vallero (Eds.), *Hazardous Waste Management* (pp. 89-115). Elsevier.

30. Jin, H. et al., (2024). Artificial intelligence in microplastic detection and pollution control. *Environmental Research, 262*, 119812.

31. Kataoka, C. & Kashiwada, S. (2021). Ecological Risks Due to Immunotoxicological Effects on Aquatic Organisms. *International Journal of Molecular Sciences,22*(1)

32. Khan, A. (2018). The Great Pacific Garbage Patch counts 1.8 trillion pieces of trash mostly plastic. *Los Angeles Times.*

33. Kumari, R. (2024). A Review of Swachh Bharat Abhiyan's Impact on India's Waste Management System. *IAHRW International Journal of Social Sciences Review*, *12*(2), 252-256.

34. Lee, Y. Y. et al., (2025). Phytoremediaton Strategies for Co-Contaminated Soils: Overcoming Challenges, Enhancing Efficiency, and Exploring Future Advancements and Innovations. *Processes, 13*(1), 132.

35. Leslie, H. A. et al., (2022). Discovery and quantification of plastic particle pollution in human blood. *Environment International*, *163*, 107199.

36. Lewis, R. C., Johns, L. E., & Meeker, J. D. (2015). Serum Biomarkers of Exposure to Perfluoroalkyl Substances in Relation to Serum Testosterone and Measures of Thyroid Function among Adults and Adolescents from NHANES 2011-2012. *International Journal of Environmental Research Public Health*, *12*(6), 6098-6114.

37. Li, A. J. et al., (2021). A review of environmental occurrence, toxicity, biotransformation and biomonitoring of volatile organic compounds. *Environmental Chemistry and Ecotoxicology*, *3*, 91-116.

38. Lusher, A. (2015). Microplastics in the Marine Environment: Distribution, Interactions and Effects. In M. Bergmann, L. Gutow, & M. Klages (Eds.), *Marine Anthropogenic Litter* (pp. 245-307). Springer International Publishing.

39. Marathe, D. et al., (2021). Persistent organic pollutants: A global issue, a global response. In *Persistent Organic Pollutants* (pp. 1-32). CRC Press.

40. Marciano, J. et al., (2024). Per/Polyfluoroalkyl Substances (PFASs) in a Marine Apex Predator (White Shark, *Carcharodon carcharias*) in the Northwest Atlantic Ocean. *ACS Environmental Au*, *4*(3), 152-161.

41. Md Meftaul, I. et al., (2020). Pesticides in the urban environment: A potential threat that knocks at the door. *Science of The Total Environment, 711*, 134612.

42. Metcalfe, C. D. et al., (2022). An introduction to the sources, fate, occurrence and effects of endocrine disrupting chemicals released into the environment. *Environmental Research, 207*, 112658.

43. O'Brien, S. et al., (2023). There's something in the air: A review of sources, prevalence and behaviour of microplastics in the atmosphere. *Science of The Total Environment, 874*, 162193.

44. Oliveira, C. S. et al., (2018). Neurodevelopmental Effects of Mercury. *Advances in Neurotoxicology, 2*, 27-86.

45. Oros, A. (2025). Bioaccumulation and Trophic Transfer of Heavy Metals in Marine Fish: Ecological and Ecosystem-Level Impacts. *Journal of Xenobiotics, 15*(2), 59.

46. Patil, A. S. et al., (2024). Drone Technology Reshaping Agriculture: A Meta-Review and Bibliometric Analysis on Fertilizer and Pesticide Deployment. *Journal of Biosystems Engineering, 49*(4), 382-398.

47. Pieper, K. J. et al., (2017). Flint Water Crisis Caused By Interrupted Corrosion Control: Investigating "Ground Zero" Home. *Environmental Science Technology, 51*(4), 2007-2014.

48. Pollutants, P. O. (2011). Stockholm Convention on Persistent Organic Pollutants. In: Geneva.

49. Qaiser, Z. et al., (2023). Microplastics in wastewaters and their potential effects on aquatic and terrestrial biota. *Case Studies in Chemical and Environmental Engineering, 8*, 100536.

50. Rasool, S. et al., (2022). A review of interactions of pesticides within various interfaces of intrinsic and organic residue amended soil environment. *Chemical Engineering Journal Advances, 11*, 100301.

51. Rauh, V. et al., (2011). Seven-year neurodevelopmental scores and prenatal exposure to chlorpyrifos, a common agricultural pesticide. *Environmental Health Perspectives, 119*(8), 1196-1201.

52. Razzaq, A. et al., (2024). Smog: Lahore needs global attention to fix it. *Environmental Challenges, 16*, 100999.

53. Ritchie, H. et al., (2023). Plastic pollution. *Our world in data.*

54. Rodríguez, A. et al., (2022). Pesticide Bioremediation: OMICs Technologies for Understanding the Processes. In S. Siddiqui, M. K. Meghvansi, & K. K. Chaudhary (Eds.), *Pesticides Bioremediation* (pp. 197-242). Springer International Publishing.

55. Ross, D. E. & Rogoff, M. J. (2012). 'What a waste…' the World Bank's call for action. *Waste Management & Research, 30*(8), 755-757.

56. Sa'adu, I. & Farsang, A. (2023). Plastic contamination in agricultural soils: a review. *Environmental Sciences Europe, 35*(1), 13.

57. Senanu, L. D. et al., (2023). The use of local materials to remove heavy metals for household-scale drinking water treatment: A review. *Environmental Technology & Innovation, 29*, 103005.

58. Smith, M. R. et al., (2022). Pollinator Deficits, Food Consumption, and Consequences for Human Health: A Modeling Study. *Environmental Health Perspectives, 130*(12), 127003.

59. Stegmann, P. et al., (2022). Plastic futures and their CO2 emissions. *Nature, 612*, 272-276.

60. Turner, A. (2021). Paint particles in the marine environment: An overlooked component of microplastics. *Water Research X, 12*, 100110.

61. UN. (2024). *Biodiversity - our strongest natural defense against climate change.*

62. UNEP. (2022a). *Clean water and sanitation.*

63. UNEP. (2022b). *Heavy metals.*

64. UNEP. (2023). *Minamata Convention COP-5 takes crucial steps in its mission of eliminating mercury pollution.*

65. Van Gerwen, M. et al., (2023). Per- and polyfluoroalkyl substances (PFAS) exposure and thyroid cancer risk. *EBioMedicine, 97*, 104831.

66. Vardhan, K. H. et al., (2019). A review on heavy metal pollution, toxicity and remedial measures: Current trends and future perspectives. *Journal of Molecular Liquids, 290*, 111197.

67. Vidu, R. et al., (2020). Removal of Heavy Metals from Wastewaters: A Challenge from Current Treatment Methods to Nanotechnology Applications. *Toxics, 8*(4), 101.

68. WHO. (2013). *Keeping our water clean: the case of water contamination in the Veneto Region, Italy.*

69. WHO. (2022). *Almost 1 million people die every year due to lead poisoning, with more children suffering long-term health effects.*

70. WHO. (2023). *Dioxins.*

71. Winiarska, E. et al, (2024). The potential impact of nano- and microplastics on human health: Understanding human health risks. *Environmental Research, 251*, 118535.

72. Yang, Y. et al., (2025). Effects of Cadmium Pollution on Human Health: A Narrative Review. *Atmosphere, 16*(2), 225.

73. Zahmatkesh, S. et al., (2024). Removing polyfluoroalkyl substances (PFAS) from wastewater with mixed matrix membranes. *Science of The Total Environment, 912*, 168881.

74. Zhao, C. et al., (2019). Thermal desorption for remediation of contaminated soil: A review. *Chemosphere, 221*, 841-855.

Chapter - 7

BIRD PARADISE OF KEOLADEO NATIONAL PARK, BHARATPUR

Dr. Neetu Nimesh[1], Varsha[2] and Dr. Sangeeta Pal[3*]

[1] Department of Zoology, Kirori Mal College, University of Delhi, Delhi-110007

[2] Department of Zoology, PDM University, Bahadurgarh, Haryana, India.

[3] Department of Zoology, Brahmanand College, Kanpur - 208 004, India.

Corresponding email: sangsaat21@gmail.com.

ABSTRACT

Keoladeo National Park in Bharatpur is a globally recognized birdwatching site, featuring a carefully managed wetland ecosystem fed by the Ajan Bund reservoir. A total of 179 bird species were recorded, including 102 residents and 77 migratory species visiting from December to March along the Central Asian Flyway (Azman et al., 2011). Birds span 18 orders and 62 families, with notable groups including waterfowl, waders, large wading birds, raptors and Passeriformes (62 species). Feeding guilds influence habitat use: herbivores prefer wetlands, frugivores occupy forests, insectivores and omnivores use both habitats, and carnivores show no specific preference. Despite their ecological and cultural significance, bird diversity is threatened by human activities and livestock grazing, with anthropogenic pressure identified as the main cause of habitat degradation affecting distribution, diversity and populations.

INTRODUCTION

Keoladeo National Park (KNP), located in Bharatpur, Rajasthan, is a protected area of global conservation significance. The park's name derives from the local deity 'Keoladeo', owing to a temple dedicated to Lord Shiva situated within its boundaries (Azeez et al., 1992). Originally a natural depression prone to seasonal flooding, the area was transformed into a managed wetland system, creating a complex of freshwater marshes that now support a spectacular diversity of life (Middleton, 1989). Encompassing 29 sq. km at the confluence of the Gambhir and Ban Ganga rivers, KNP is a lush haven for aquatic, semi-aquatic and terrestrial wildlife (Azeez et al., 1992).

Recognized for its outstanding universal value, KNP was declared a Ramsar site in 1981 and inscribed as a UNESCO World Heritage Site in 1985. It is internationally renowned as an "ornithologist's paradise", providing habitat for over 400 bird species, including numerous migrants that travel the Central Asian Flyway (Bhadouria et al., 2012). The park is a crucial staging and wintering ground for species ranging from waterfowl like the Siberian crane to large nesting colonies of storks, ibises and herons.

OBJECTIVE OF THE STUDY

A. To analyze the diversity of bird species in Keoladeo National Park.

B. To evaluate the species richness of avifauna specifically during the winter months.

RATIONALE OF THE STUDY

KNP's wetlands, woodlands, and grasslands support many resident and migratory birds, but tourism, deforestation, urbanization, and human activities threaten the ecosystem. Studying bird communities in winter helps establish a baseline for avifaunal diversity and supports future research, conservation, and management efforts.

AVIFAUNAL CHECKLIST

KNP is rightly known as a "Bird Paradise" due to the rich avifaunal assemblage it supports (Bhadouria et al., 2012). It functions as a critical staging ground for migratory waterfowl arriving via the Central Asian Flyway and hosts one of the world's most spectacular heronries, with 15 species nesting in large numbers. The park's avifauna can be categorized as follows:

- **Waterbirds:** This group includes species such as pelicans, herons, storks, ibises, cormorants, darters, gadwall, shoveler, common teal, ruff, sandpipers, spoonbill, and the Sarus crane (*Grus antigone*) (Archibald et al., 2003).
- **Landbirds:** This category includes bee-eaters, bulbuls, grey francolin, quails, hornbills, eagles, warblers and babblers.
- **Resident Birds:** Resident species include the openbill stork, painted stork, black-necked stork, woolly-necked stork and the regularly breeding dusky horned owl.
- **Migratory Birds:** Peak winter brings an influx of migratory ducks such as the northern pintail, common teal, gadwall, graylag, bar-headed goose, wigeon and common pochard. Migratory raptors like harriers and eagles also dominate during this season.
- **Resident-Migratory Birds:** Some species, like the garganey, water rail and Eurasian eagle owl, exhibit both resident and migratory populations within the park.

LITERATURE REVIEW

Bird survival in urban and semi-urban environments depends on adaptability and resource availability. Modern planning increasingly incorporates wetlands and green spaces to support bird diversity. KNP provides an important habitat for resident and migratory birds and acts as a bioindicator of ecosystem health. However, anthropogenic pressures such as noise, light, pollution, and urban expansion can negatively affect bird reproduction and survival (Patra et al., 2017; Upadhyay et al., 2014).

Hydrology and biomass management are essential for maintaining wetland health, with trophic assessment and nutrient control being key strategies (Saluja & Garg, 2017; Reddy & Char, 2006). Studies have documented avian diversity, population trends, and monitoring methods in the park (Bhadouria et al., 2019; Urfi et al., 2005) Species-focused research highlights the conservation needs of the Sarus Crane (Kulshreshtha et al., 2013; Archibald et al., 2003; Borad et al., 2002) and the Indian Peafowl (Dookia, 2015).

Major threats include pesticide use, water scarcity, invasive species, urbanization, and altered hydrology (Tiwari & Upadhyay, 2017; Neeraj et al., 2016; Bhadouria et al., 2012; Pain et al., 2004; Muralidharan, 1993). Socio-economic studies show that local communities depend on the park, and regulated activities like controlled grazing can support biodiversity (Chauhan & Gopal, 2001a; Azeez et al., 1992). Technological tools such as Synthetic Aperture Radar and satellite imagery also aid wetland monitoring and conservation planning (Choudhary et al., 2021; Srivastava et al., 2009, 2007). Overall, integrated ecological, social and technological approaches are essential for sustaining biodiversity in the park.

MATERIALS AND METHODS

The study was conducted in KNP (29 km^2; 27°8′–27°12′ N, 77°30′–77°34′ E), a UNESCO World Heritage Site located in a semi-arid plain. The area experiences temperatures from 5–47 °C with an average annual rainfall of about 662 mm (Bhadouria et al., 2019; Sharma et al., 2015).

Avifauna Survey: The survey was conducted from December 2022 to February 2023 across forests, grasslands, and wetlands, recording 179 bird species. Observations focused on ecological and behavioral patterns, and photographs were taken using Canon EOS 700D and Canon EOS 90D cameras with Sigma lenses.

OBSERVATION

S.N	Species with scientific name	Habitat	System	IUCN status	Population trend	Status	Dec.	Jan.	Feb	Mar.
	Order- Anseriforme, Family- Anatidae									
1.	Lesser-whistling duck *Dendrocygna javanica*	F	T, FW, MW	LC	+	R, C	15621	21023	12189	4744
2.	Bar-headed goose *Anser indicus*	W	T, FW	LC	-	M, VC	5703	11883	18445	9541
3.	Greylag goose *Anser anser*	G	T, FW	LC	+	M, VC	12479	17463	16888	280
4.	Ruddy shelduck *Tadorna ferruginea*	G	T, FW	LC	?	LM, SN	954	1312	2117	1068
5.	Red-crested pochard *Netta rufina*	W	T, FW	LC	?	M, C	1535	2411	1747	185
6.	Garganey *Spatula querquedula*	G	T, FW, MW	LC	-	PM	945	935	1575	2951

7.	Northern shoveler *Spatula clypeata*	G	T, FW, MW	LC	-	M, VC	9627	15241	22717	7010
8.	Gadwall *Maraca strepera*	W	T, FW	LC	+	M, C	7167	13539	12166	2290
9.	Eurasian wigeon *Maraca penelope*	G	T, FW, MW	LC	-	M, C	1886	2788	2472	657
10.	Indian spot-billed duck *Anas poecilorhynchos*	W	T, FW	LC	-	R, C	5431	6649	6968	2236
11.	Mallard *Anas platyrhyncha*	W	T, FW	LC	+	M, U	504	106	179	13
12.	Northern pintail *Anas acuta*	W	T, FW, MW	LC	-	M, VC	22596	28794	31214	1209
13.	Common teal *Anas crecca*	F	T, FW	LC	?	M, C	8161	13983	22389	3945
	Order- Galliformes, Family- Phasianidae									
14.	Indian peafowl *Pavo cristatus*	F	T	LC	=	R, C	1539	1461	2156	1180
15.	Black francolin *Francolinus francolinus*	G	T	LC	=	R, B	18	12	15	6

16.	Grey francolin *Francolinus pondicerianus*	G	T	LC	=	R, B, VC	1169	1270	2006	847
	Order- Phoenicopteriformes, Family- Phoenicopteridae									
17.	Greater flamingo *Phoenicopterus roseus*	W	T, FW, MW	LC	+		173	181	354	178
	Order- Phoenicopteriforme, Family- Podicipedidae									
18.	Little grebe *Tachybaptus ruficollis*	W	T, FW, MW	LC	-	R, C	1906	2335	2952	765
19.	Great- crested grebe *Podiceps cristatus*	W	T, FW, MW	LC	=	M, O	57	22	59	2
	Order- Columbiformes, Family- Columbidae									
20.	Rock dove *Columba livia*	RO	T	LC	-	R, C	1448	1884	2203	1633
21.	Laughter dove *Streptopelia senegalensis*	G	T, FW	LC	=	R, C	1398	1124	3069	1185
22.	Yellow-footed green pigeon *Treron Phoenicopterus*	F	T	LC	+	R, C	4121	6043	3314	697

	Order- Pterocliformes, Family- Pteroclidae									
23.	Chestnut-bellied sandgrouse *Pterocles exustus*	G	T	LC	=	LM, U	5	3	37	7
	Order- Caprimulgiformes, Family- Caprimulgidae									
24.	Grey nightjar *Caprimulgus indicus*	F	T	LC	=	R, O	61	31	19	2
25.	Large-tailed nightjar *Caprimulgus macrurus*	F	T, FW	LC	=		20	37	47	3
	Order- Caprimulgiformes, Family- Apodidae									
26.	Little swift *Apus affinis*	F	T	LC	+	R, C	1559	1088	1801	340
	Order- Cuculiformes, Family- Cuculidae									
27.	Greater coucal *Centropus sinensis*	F	T, FW	LC	=	R, C	604	1109	989	441
28.	Jacobin cuckoo *Clamator jacobinus*	G	T	LC	=	M, B, C	2	1	1	0
29.	Common koel *Eudynamys scolopaceus*	F	T	LC	=	R, O	81	84	62	74
30.	Common hawk cuckoo *Hierococcyx varius*	F	T	LC	=	R, C	81	48	68	65

	Order- Gruiformes, Family- Rallidae									
31.	Western water rail *Rallus aquaticus*	G	T, FW	LC	-	LM, O	4	8	9	0
32.	White-breasted waterhen *Amaurornis phoenicurus*	F	T, FW	LC	?	R, U	2034	2459	2080	1123
33.	Watercock *Gallicrex cinerea*	W	T, FW	LC	-	B	3	3	6	1
34.	Common moorhen *Gallinula chloropus*	W	T, FW	LC	=	R, M, C	8767	10563	16251	5351
35.	Common coot *Fulica atra*	G	T, FW, MW	LC	+	VC, M	32911	35260	69820	10028
	Order- Gruiformes, Family- Gruidae									
36.	Sarus crane *Antigone antigone*	G	T, FW	VU	-	R, C, B, H, WM	895	865	1889	883
37.	Common crane *Grus grus*	G	T, FW, MW	LC	+	M, O	561	298	1937	1576
	Order- Pelecaniformes, Family- Ciconiidae									
38.	Painted stork *Mycteria leucocephala*	W	T, FW, MW	NT	-	R, C, BL	36593	22676	21581	3937

39.	Asian openbill *Anastomus oscitans*	W	T, FW	LC	?	R, C, B	1594	431	604	181
40.	Wolly-necked stork *Ciconia episcopus*	G	T, FW, MW	NT	-	R, C, B	399	251	277	168
41.	European white stork *Ciconia ciconia*	G	T, FW	LC	+	M, U	4	2	42	14
42.	Black necked stork *Ephippiohynchus asiaticus*	W	T, FW, MW	NT	-	R, C	351	491	652	267
	Order- Pelecaniformes, Family- Pelecanidae									
43.	Great white pelican *Pelecanus onocrotalus*	W	T, FW, MW	LC	?	LM, O	5418	7204	4105	1348
44.	Dalmatian pelican *Pelecanus crispus*	W	T, FW, MW	NT	-	M, O	243	555	608	187
	Order- Pelecaniformes, Family- Ardeidae									
45.	Yellow bittern *Ixobrychus sinensis*	F	T, FW	LC	?	LM, U	23	55	74	19
46.	Cinnamon bittern *Ixobrychus cinnamomeus*	F	T, FW	LC	=	LM, U	3	7	12	3

47.	Black bittern *Ixobrychus flavicollis*	F	T, FW, MW	LC	-	LM, O	149	216	277	104
48.	Black-crowned night heron *Nycticorax nycticorax*	F	T, FW, MW	LC	-	R, C	1847	2800	3135	929
49.	Straited heron *Butorides striata*	W	T, FW, MW	LC	=	R, U	179	197	226	93
50.	Indian pond heron *Ardeola grayii*	F	T, FW, MW	LC	?	R, C	3141	2656	4385	1642
51.	Cattle egret *Bubulcus ibis*	G, W	T, FW	LC	+	R, C	1872	3104	2015	1108
52.	Grey heron *Ardea cinerea*	F	T, FW, MW	LC	?	R, C	2058	2091	4441	1615
53.	Purple heron *Ardea purpurea*	W	T, FW, MW	LC	-	R, C	2041	1901	2663	1008
54.	Great egret *Ardea alba*	G, W	T, FW, MW	LC	?	R, C	1688	1864	3406	1957
55	Intermediate erget *Ardea intermedia*	F	T, FW, MW	LC	-	R, C	2259	2092	4594	1502
56.	Little egret *Egretta garzetta*	G	T, FW	LC	+	R, C	1459	1894	2569	1156

	Order- Pelecaniformes, Family- Threskiornithidae									
57.	Black-headed ibis *Threskiornis melanocephalus*	F	T, FW, MW	NT	-	R, C	3500	2867	4012	1942
58.	Eurasian spoonbill *Platalea leucorodia*	F	T, FW, MW	LC	?	R, C, O	3530	3869	5489	1775
59.	Indian black ibis *Pseudibis papillosa*	G	T, FW	LC	-	LM, O	57	112	268	113
60.	Glossy ibis *Plegadis falcinellus*	W	T, FW, MW	LC	-	LM, C	2091	2843	3883	2006
	Order- Pelecaniformes, Family- Phalacrocoracidae									
61.	Great cormorant *Phalacrocorax carbo*	F, W	T, FW, MW	LC	+	R, C	4343	5286	6793	780
62.	Indian cormorant *Phalacrocorax fuscicollis*	F	T, FW, MW	LC	?	R, C	2787	2537	2024	1435
	Order- Pelecaniformes, Family- Anhingidae									
63.	Oriental darter *Anhinga melanogaster*	F	T, FW, MW	NT	-	R, C	3909	2911	3421	1547

	Order- Charadriiformes, Family- Recurvirostridae									
64.	Black winged stilt *Himantopus himantopus*	G	T, FW, MW	LC	+	LM, C	2412	2257	5009	2331
	Order- Charadriiformes, Family- Charadriidae									
65.	Little ringed plover *Charadrius dubius*	F	T, FW, MW	LC	=	M, O	18	19	123	83
66.	Kentish plover *Charadrius alexandrinus*	G	T, FW, MW	LC	-	?	10	16	17	42
67.	Northern lapwing *Vanellus vanellus*	G	T, MW	NT	-	M, O	25	17	45	4
68	Yellow-wattled lapwing *Vanellus malabaricus*	G	T, FW	LC	=	R, O	76	127	280	85
69	Grey-headed lapwing *Vanilla cinereus*	G	T, FW	LC	-	M, U	2	11	28	1
70	Red-wattled lapwing *Vanellus indicus*	G	T, FW	LC	?	R, C	1866	1221	4441	2187
71.	White-tailed lapwing *Vanellus leucurus*	W	T, FW, MW	LC	?	M, C	446	840	1384	334

	Order- Charadriiformes, Family- Rostratulidae									
72.	Greater painted-snipe *Rostratula benghalensis*	G	T, FW, MW	LC	-	R, O	50	54	71	33
	Order- Charadriiformes, Family- Jacanidae									
73.	Bronze-winged jacana *Metopidius indicus*	G	T, FW	LC	?	R, C	1432	1110	1807	715
	Order- Charadriiformes, Family- Scolopacidae									
74.	Eurasian curlew *Numenius arquata*	F	T, FW, MW	NT	-	M, U	15	34	17	4
75.	Black-tailed godwit *Limosa limosa*	G	T, FW, MW	NT	-	M, U	197	427	996	209
76.	Ruff *Calidris canutus*	G	T, FW, MW	LC	-	M, O	1082	978	2635	794
77.	Temminck's stint *Calidris temminckii*	G	T, FW, MW	LC	?	M, O	113	574	350	239
78.	Green sandpiper *Tringa ocropus*	F	T, FW	LC	+	M, C	618	421	887	332
79.	Wood sandpiper *Tringa glareora*	F	T, FW	LC	=	M, C	1135	1079	2510	1622
80.	Marsh sandpiper *Tringa stagnatilis*	W	T, FW, MW	LC	-	M, O	101	91	189	102

	Order- Charadriiformes, Family- Glareolidae									
81.	Indian courser *Cursorius coromandelicus*	G	T	LC	=	R, C	46	44	137	48
	Order- Charadriiformes, Family- Laridae									
82.	Black-headed gull *Chroicocephalus ridibundus*	G	T, FW, MW	LC	?	M, U	18	10	23	10
83.	Pallas's gull *Ichthyaetus icthyaetus*	W	T, FW, MW	LC	+		1	1	7	3
84.	Black-bellied tern *Sterna acuticauda*	W	T, FW	EN	-	HA	274	214	205	55
	Order- Accipitriformes, Family- Pandionidae									
85.	Osprey *Pandion haliaetus*	F	T, FW, MW	LC	+	M, O	27	20	15	3
	Order- Accipitriformes, Family- Accipitridae									
86.	Black-winged kite *Elanus caeruleus*	G	T	LC	=	R, C	17	4	0	0
87.	Oriental honey buzzard *Pernis ptilorhynchus*	F	T	LC	-	R, C	76	75	129	54
88	Egyptian vulture *Neophron percnopterus*	RO	T, FW	EN	-	R, C	870	868	1238	600

89.	Crested serpent eagle *Spilornis cheela*	F	T, FW	LC	=	LM, C	256	329	313	82
90.	Short-toed eagle *Circaetus gallicus*	G	T	LC	=	R, C	11	32	25	6
91.	Greater spotted eagle *Clanga clanga*	G, W	T, FW	VU	-		636	768	905	284
92.	Eastern imperial eagle *Auila heliaca*	F, W	T, FW	VU	-	M, C	158	232	252	70
93.	Montagu's harrier *Circus pygargus*	G	T, FW	LC	-	M, O	2	5	7	2
94.	Shikra *Accipiter badius*	F	T	LC	=	R, C	185	253	242	111
95.	Brahminy kite *Haliastur indicus*	F	T, FW, MW	LC	-	LM, ?, U	15	8	12	17
96.	Black kite *Milvus migrans*	F	T, FW, MW	LC	=	R, U	372	490	1097	405
	Order- Strigiformes, Family- Tytonidae									
97.	Common barn owl *Tyto alba*	G	T, FW	LC	=	M, O	2	6	11	1

	Order- Strigiformes, Family- Strigidae									
98.	Brown boobook *Ninox scutulata*	F	T	LC	-	LM, U	10	12	26	1
99.	Spotted owlet *Athene brama*	T	T	LC	=	R, C	751	869	959	345
100.	Collared scops owl *Otus bakkamoena*	F	T, FW	LC	=	R, O	263	869	959	345
101.	Short-eared owl *Asio flammeus*	G	T, FW	LC	-	M, U	1	2	3	2
102.	Dusky eagle owl *Bubo coromandus*	F	T, FW	LC	-	R, C	268	376	694	216
103.	Brown fish owl *Ketupa zeylonensis*	F	T, FW	LC	-	LM, U	3	2	8	0
	Order- Bucerotiformes, Family- Bucerotidae									
104.	Indian grey hornbill *Ocyceros birostris*	F	T	LC	=	R, B	359	432	775	468
	Order- Bucerotiformes, Family- Upupidae									
105.	Common hoopoe *Upupa epops*	G	T	LC	-	R, M	400	345	505	141

	Order- Piciformes, Family- Picidae									
106.	Black-rumped woodpecker *Dinopium benghalense*	F	T	LC	=	R, B	293	337	378	175
107.	Yellow-crowned woodpecker *Dendrocopos mahrattensis*	F	T, FW	LC	=	R	79	139	172	63
	Order- Piciformes, Family- Ramphastidae									
108.	Brown headed barbet *Psilopogon zeylanicus*	F	T	LC	=	R, N	175	178	294	231
109.	Coppersmith barbet *Psilopogon haemacephalus*	F	T	LC	+	R, B	175	168	325	181
	Order- Coraciformes, Family- Meropidae									
110.	Green bee-eater *Merops orientalis*	F	T, FW	LC	+	R, C	713	368	1484	1723
	Order- Coraciiformes, Family- Coraciidae									
111.	Indian roller *Coracias benghalensis*	F	T	LC	+	R, C	199	322	538	180

	Order- Coraciiformes, Family- Alcedinidae									
112.	Common kingfisher *Alcedo atthis*	W	T, FW, MW	LC	?	C, R	491	572	589	151
113.	Pied kingfisher *Ceryle rudis*	F	T, FW, MW	LC	?	R, C	487	446	597	209
114.	White-throated kingfisher *Halcyon smyrnensis*	F	T, FW, MW	LC	+	R, VC	1698	1713	2330	1050
	Order- Falconiformes, Family- Falconidae									
115.	Common kestrel *Falco tinnunculus*	F	T	LC	-	LM, O	47	59	67	20
116.	Peregrine falcon *Falco peregrinus*	RO	T, FW, MW	LC	+	M, O	23	16	26	9
	Order- Psittaciformes, Family- Psittaculidae									
117.	Plum-headed parakeet *Psittacula cyanocephala*	F	T	LC	-	LM, O	89	29	74	62
118.	Rose-ringed parakeet *Psittacula krameri*	F	T, FW	LC	+	R, C	6792	11536	11301	3636
	Order- Passeriformes, Family- Campephagidae									
119.	Small minivet *Pericrocotus brevirostris*	F	T	LC	=	R, C	231	178	328	56

120.	Long-tailed minivet *Pericrocotus flammeus*	F	T	LC	-		33	60	43	1
	Order- Passeriformes, Family- Dicruridae									
121.	Black drongo *Dicrurus macrocercus*	G	T	LC	?	R, C	1388	1108	3206	1351
	Order- Passeriformes, Family- Rhipiduridae									
122.	White-throated fantail *Rhipidura albicollis*	F	T	LC	=	?, HA	56	56	53	10
	Order- Passeriformes, Family- Laniidae									
123.	Brown shrike *Lanius cristatus*	F	T	LC	-	M, U	3	2	16	3
124.	Isabelline shrike *Lanius isabbellinus*	G	T, FW	LC	=	M, O	37	22	37	2
125.	Bay-backed shrike *Lanius vittatus*	RO	T	LC	=	R, C	282	314	286	102
126.	Long-tailed shrike *Lanius schach*	F	T, FW	LC	?	LM, C	453	479	481	112
127.	Great grey shrike *Lanius excubitor*	F	T, FW	LC	-		55	35	120	4
	Order- Passeriformes, Family- Corvidae									
128.	Rufous treepie *Dendrocitta vagabunda*	F	T	LC	-	R, C	950	1052	1535	577

139.	House crow *Corvus splendens*	T	T	LC	=	R, C	1917	1945	4926	1460
	Order- Passeriformes, Family- Monarchidae									
130.	Indian paradise-flycatcher *Terpsiphone paradisi*	F	T	LC	=	R, M, C	5	7	11	31
	Order- Passeriformes, Family- Nectariniidae									
131.	Purple sunbird *Cinnyris asiaticus*	F	T, FW	LC	=	R, C	719	476	1077	576
	Order- Passeriformes, Family- Ploceidae									
132.	Baya weaver *Ploceus philippinus*	F	T, FW	LC	=	R, C	170	155	68	39
	Order- Passeriformes, Family- Estrildidae									
133.	Red avadavat *Amandava amandava*	F	T, FW	LC	=	R, C	252	439	292	25
134.	Indian silverbill *Euodice malabarica*	F	T	LC	=	R, C	600	728	729	159
135.	Scaly-breasted munia *Lonchura punctulata*	F	T	LC	=	R, O	54	40	41	20

	Order- Passeriformes, Family- Passeridae									
136.	House sparrow *Passer domesticus*	F	T, FW	LC	-	M, O	866	1609	1912	919
	Order- Passeriformes, Family- Motacillidae									
137.	Paddyfield pipit *Anthus rufulus*	G	T	LC	=	R, C	97	66	189	95
138.	Grey wagtail *Motacilla cinerea*	W	T, FW	LC	=	M, O	247	200	221	94
139.	Citrine wagtail *Motacilla citreola*	G	T, FW, MW	LC	+	M, C	997	1129	1650	286
140.	White wagtail *Motacilla alba*	G	T, FW	LC	=	M, C	256	450	487	78
	Order- Passeriformes, Family- Fringillidae									
141.	Common rosefinch *Erythrina erthrina*	F	T, FW	LC	-	LM, O	10	3	4	50
	Order- Passeriformes, Family- Stenostitidae									
142.	Grey-headed canary-flycatcher *Culicicapa ceylonensis*	F	T	LC	=		155	1	181	16

	Order- Passeriformes, Family- Alaudidae									
143.	Rufous-tailed lark *Ammomanes phoenicura*	G	T	LC	=	R, O, ?	34	35	86	16
144.	Ashy-crowned sparrow lark *Eremopterix griseus*	G	T	LC	=	LM, O	66	24	140	26
145.	Crested lark *Galerida cristata*	G	T, MW	LC	-	R, C	34	44	105	13
	Order- Passeriformes, Family- Cisticolidae									
145.	Grey-breasted prinia *Prinia hodgsonii*	F	T	LC	=		26	6	31	0
147.	Ashy prinia *Prinia socialis*	F	T, FW	LC	=	R, O	561	839	741	329
143.	Plain prinia *Prinia inornata*	G	T, FW, MW	LC	=	R, C	828	849	837	178
149.	Common tailorbird *Orthotomus sutorius*	F	T	LC	=	R, C	321	415	670	313
	Order- Passeriformes, Family- Acrocephalidae									
15[illegible].	Blyth's reed warbler *Acrocephalus dumetorum*	F	T, FW	LC	+	LM, O	245	343	251	85

	Order- Passeriformes, Family-Hirundinidae									
151.	Streak-throated swallow *Petrochelidon fluvicoia*	G	T, FW	LC	+		106	133	306	45
152.	Wire-tailed swallow *Hirundo smithii*	G	T, FW	LC	+	R, C	633	545	1009	350
	Order- Passeriformes, Family-Pycnonotidae									
153.	Red-whiskered bulbul *Pycnonotus jocosus*	F	T, FW	LC	-	R, C	35	17	29	4
154.	White-eared bulbul *Pycnonotus leucotis*	F, W	T, FW	LC	-	R, C	1684	1553	1969	862
155.	Red-vented bulbul *Pycnonotus luteolus*	F	T	LC	+	R, C	2724	2197	3827	1714
	Order- Passeriformes, Family-Phylloscopidae									
156.	Common chiffcgaff *Phylloscopus collybita*	F	T	LC	+	M, C	1002	600	483	172
	Order- Passeriformes, Family- Sylviidae									
157.	Lesser whitethroat *Curruca curruca*	F	T, FW	LC	=	M, C	1037	871	988	326

	Order- Passeriformes, Family- Zosteropidae									
158.	Oriental white-eye *Zosterops palpebrosus*	F	T	LC	-	R, C	207	249	360	176
	Order- Passeriformes, Family- Leiothrichidae									
159.	Large grey babbler *Argya malcolmi*	T	T	LC	=	R, C	653	939	1174	375
160.	Jungle babbler *Turdoides striata*	F	T	LC	=	R, C	3608	4204	5690	1854
	Order- Passeriformes, Family- Sturnidae									
161.	Common starling *Sturnus vulgaris*	F	T, MW	LC	-	M, O	65	155	607	4
162.	Rosy starling *Pastor roseus*	G	T	LC	?	M, O	440	389	46	1789
163.	Brahminy starling *Sturnia pagodarum*	F	T	LC	?	R, C	3220	2195	3345	1645
164.	Common myna *Acridotheres tristis*	F	T	LC	+	R, C	1889	2266	3649	1186
165.	Bank myna *Acridotheres ginginianus*	T	T, FW	LC	+	R, O	3382	3458	3949	253

	Order- Passeriformes, Family- Muscicapidae									
166.	Indian robin *Saxicoloides fulicatus*	RO	T,	LC	=	R, C	767	716	1140	437
167.	Oriental magpie robin *Copsychus saularis*	F	T, FW	LC	=	R, C	1495	1288	1713	684
168.	Tickell's blue flycatcher *Cyornis tickelliae*	F	T,	LC	=	R, C	11	11	5	5
169.	Asian verditer flycatcher *Eumyias thalassinus*	F	T,	LC	=	M, O	11	10	6	4
170.	Bluethroat *Luscinia svecica*	F	T, FW, MW	LC	=		821	875	884	152
171.	Red-breasted flycatcher *Ficedula parva*	F	T,	LC	+	M, U	404	493	478	172
172.	Taiga flycatcher *Ficedula albicilla*	F	T,	LC	=	M, C	27	60	45	7
173.	Ultramarine catcher *Ficedula superciliaris*	F	T,	LC	=	M, U	3	3	5	7
174.	Black redstart *Phoenicurus ochruros*	RO, T	T,	LC	+	M, C	456	472	505	157

175	Pied bush chat *Saxicola caprata*	W	T, F	LC	=	R, C	523	488	817	264
176	Grey bush chat *Saxicola ferreus*	F	T, F	LC	=	LM, U	5	5	8	2
177	Desert wheatear *Oenanthe deserti*	RO	T,	LC	=	LM, O	16	6	12	2
178	Brown rock chat *Oenanthe fusca*	RO	T,	LC	=	R, C	76	73	88	50
	Order- Passeriformes, Family- Turdidae									
179	Orange-headed thrushush *Geokichla citrina*	F	T, F	LC	-	M, O	82	102	81	11

Table-1: R-Resident, C-Common, VC-Very common, M-Migratory, O-Occasional, LM-Local migrant, SN-Small no, PM-Passage migrant, U-Uncommon, B-Breeds WM-Winter migrant, H-High site fidelity, BL-Breed in large no, ?-Unknown, FW-Freshwater (inland), MW-Marine, F- Forest, RO-Rocky area, T-Terrestrial, G-Grassland, W-Wetland, - Decreasing, + Increasing, = Stable, LC- Least Concern.

A. **Diversity of birds**

Birds in KNP belong to 18 orders and 62 families. The IUCN Red Data Book maintains records of globally threatened species (Neeraj, et al., 2016).

B. **Species richness of avifauna**

In the study area 179 bird species were identified of which 102 are Residential, 77 are Migrant are recorded under the 18 orders and 62 families. The data can be collected by ebird (*Bar Charts - eBird*, no date) which providing information regarding distribution and abundance of bird.

Discussion

Pictures can be deceitful, and you will always wonder how small a spotted owl[99] is when you see it. About the size of common myna[164], because they have heart-shaped spots. Pallas gull[83] also has a less common name, the greater black-headed gull[82]. The body of greater cormorant[61] is clearly larger than that of the Indian cormorant[62]. it looked stout and had a whitish head and was black all over. One of the most beautiful birds is the plum headed parakeet[117], it is a delight to see that these are smaller than the commonly seen rose-ringed parakeet[118]. One of the most popular birds ospreys[85]. Yellow-footed green pigeon[22] was one of the most colourful birds. These same colours help them camouflage it is a different subspecies. Red-crested pochard[05] is migratory birdduring winters. Brown fish owl[103] is the one who hides comfortably under the shadows of the forest canopy. Collared scops owl[100] found on forest edge and identified by its dark eyes, facial disc. Short-eared owl[101] is cold seasoned inhabitant. Brown Boobook[98] (Hawk owl) is nocturnal bird activate at dusk and dawn. Dalmatian pelican[44] isn't as gregarious and is usually seen feeding or in small flocks, usually 4-6 not larger than other pelicans. Their status falls under vulnerable, with habitat loss, poaching and silting affecting them most. Common coots[35] have noisy flights because they run on water before taking flight. Crested serpent eagle[89] can be found in wooded regions with some kind of water source. Blyth's reed warbler[150] is found on trees and vegetation close to water sources. Indian roller[111] pictures make them seem smaller than they are by their size. They are the size of pigeons, Rock dove[20] and are called Neelkanth. Marsh sandpipers[80] are not larger than green sandpipers[78] as green sandpipers[78] do appear taller due to their slender forms. Wood sandpiper[77] is more profusely spotted than green sandpiper. They have a more uniform spotting pattern. Black necked stork[42] is seen in pairs near the wetland. Mallard[11] is migratory bird from Central Asia. Asian verditer flycatcher[169] as its name suggests, is a certain shade of blue. Eastern imperial eagle[92] looks similar to a black kite[96] and it is called black kite eagle[96]. Egyptian vulture[88] founds in open and nests in open area on cliffs. Brahminy kite[95] is significant in park. Red avadavat[133] is the most eye catching munia in breeding plume and is seen most frequently. Indian grey hornbill[104] are seen in pairs outside nesting period will often see them glide a short distance from a higher perch to a lower one. Purple sunbird[131], a fair few eclipse males in non-breeding plumes. Common hoopoe[105] are lazy nesters, they choose existing hollows as nests. If predators come to their nest, the chick excretes and forces the predator to exit. When it flies, it has stripes like Zebra. Greater spotted eagle[91] juveniles look more beautiful. The spots on them fade after becoming adult. They prey on both waterfowl and fish. Red-whiskered bulbul[153] is most likely fleeing at once and will observe us curiously, as one can be seen doing. Coppersmith barbet[109] is named because of the bristle like feather on its beak. It is a very colourful bird and is more frequently seen. Brown headed barbet[108] is median sized green and brown tree dwelling bird. Shikra[94] is the most widespread and easily seen bird. These are very alert birds and a delight to observe as they fly between trees. Little grebe[18] is a lot more alert than average water birds. It is fun to watch a grebe disappear, for they dive almost without a ripple. Great-crested grebe[19] is diving bird feeds on fish and crustacean. Eurasian spoonbill's[58] beak looks flat when seen from sideways. Also notice their relative size in comparison to a Great egret[54]. Oriental darters[63] are expert hunters and catch fish almost every time and they pierce fish to kill them with their beaks. Sarus cranes[36] had fresh feathers and were seen alone the two times Unlike most Common kingfishers[112] that dive from perches, pied kingfisher hovers[113] above water before diving. Indian pond heron[50] can see with its body more upright than usual while striking a fish.

Common kestrel[115] is distinctive hovery flight while hunting. Grey francolins[16] aren't exactly grey birds. They aren't frequently seen, despite their alertness as once frequent game birds. These can be seen as about 6-8 adults and subadults walking on a wall. Black francolins[15] is hard to see but due to its distinctive metallic cal it is recognized. Common rose finch[141] is fairly common towards the end of migration season, from January to April. Asian openbill[39] seem more like an egret or a heron at a distance. The black outer edges of its wings are useful in identifying this one in flight. Grey nightjar[24] is most common and widespread in park. Large-tailed nightjar[25] is nocturnal with brown, grey and black mettled plumage. Greater flamingo[17] are birds that have been seeing for a while in Keoladeo National Park. They appear as a few white dots, and sometimes seen them fly as well. Western water rail[31] is very rare species and difficult to find in park. Streak throated swallow[151] was convinced that this was a martin at first. Usually found near water bodies and canals. Indian courser[81] is a bird with varying shades of brown, it blends with its surroundings and can be hard to see if it does move at all. Black crowned night heron[48] can be found asleep between the branches of trees. Indian robin1[66] can be found in dry, scrub habitats. Can be found with drooping wings and cocking tails in fashion. Small minivet[119] occupies the foliage of the tree canopy area, sticking close to the edge of wooded area. Long-tailed minivet[120] is beautiful bright red colour bird. White throated fantail[122] is also documentd in park. Black drongo[121] is distinctive forked tail and known for its aggressive behaviour. Orange headed thrush[179] is hard to miss, but we can easily miss it. They usually like to stay hidden, flying at the slightest movement. If caught, they tend to freeze and wait. Lesser whitethroat[157] can see hopping as they glance back. They hunts insects. They make a call similar to the sound of humans clicking their tongues. Paddy field pipit[137] flight resembles that of wagtails, in a bobbing fashion. Grey headed canary flycatcher[142] is hard to miss. Yet It can easily disappear. Indian peacocks[14] can be seen fairly easily across India. Black winged kite[86], commonly called the black shouldered kite, about the size of a common crow. Black rumpled woodpecker[106] is a shy bird, by approaching them, they quickly climb or fly. Yellow wattled lapwing[68] is an alert bird and walks away at the slightest chance. Common kingfisher[112] is a House sparrow[136] sized kingfisher, White throated kingfisher[114] is almost 1.5 times its size. They startle a lot more frequently. Indian paradise flycatcher[130] is gorgeous, restless bird. in flight look like a gymnast's ribbon, mainly for its tail. Yellow crowned woodpecker[107] is identified by its spotted pied back and striking yellow crown. When they hammer decaying bark for insects, it sounds 'pecking'. Great white pelican[43] has swimming and feeding rituals like a slow ritual or dance. The colour of Desert wheatear[177] is almost entirely to the soil behind it. Rosy starlings[162] can be seen in flocks of 50-100 to larger groups of 200-500. Common starling[161] is very common in park. Green bee eater[110] was sitting under a shaft of light with the shadow of a tree in the background. Gadwall[08] is a duck that doesn't stand out. Mainly due to its grey shades and extensive ripple details that make it harder to see at a distance. Dusky eagle owl[102] is named for their sheer size, large owls, about the height of a black kite but much stockier. Rufous tailed larks[143] fly during breeding in a climb and dive loop similar to that Ashy crowned sparrow lark[144]. Crested lark[145] is found in grassy areas. When the Green sandpiper[78] starts bobbing its tail like a wagtail, try and move a few steps away from it, as they do that when nervous. Bar-headed goose[02] will fly at averaging from 11000 ft to almost 24000 ft as they journey back. Great Grey shrike's tail[127] in flight shows white edges and white on the wings is more prominent. Isabelline shrike[124] is pale sandy coloured bird common in dry and bushy areas. Black headed gull[82] is one of the smaller gulls, similar in size to the common crow. Northern pintail[12] in sufficiently deep water, can regularly be seen 'upending', where they tilt perpendicularly and feed while their tails stick out of the water. Large grey babbler[159] eyes give it a constantly angry expression, Perfect angry birds. Wolly necked stork[40] are mostly preferred where water levels is relatively low and are known by their feathered neck which appear woolly. Scaly breasted munia[135] are as sweet as they seem, they eat voraciously and will keep pushing the others away to get access to more grains. Ashy crowned sparrow lark[144] has ashy brown colours that mix well with the hues of dry scrubland that are most often found. Eurasian curlew[74] were once popular game birds and are thus extremely wary of human presence. They need to be approached carefully and slowly and then almost always fly if one moves too close. Common chiffchaff warbler[156] is one of the more fun birds to spot. They won't sit still for us to look for characteristics; they will twist, sit between branches and more often than not, vanish. Temminck's stint[77] can be confusing, even after seeing them for a while. Bay backed shrike[125] is the smallest resident Indian shrike. Cranes and storks are easily the largest birds. Measuring an average of 140cm from head to toe, the common crane[37], Europian white stork[41] is graceful for its size. Common teal[13] has spot like markings on its front and a ripple pattern at the edge of its wings that aid in its camouflage. Peregrine falcon[116] uses gravity to accelerate to speeds of up to 350-390 kph. Adaptation it has a respiratory system which is a significant reason for its speed. Black kite[96] isn't very powerful. While it can hunt lizards, worms, and sometimes even injured birds, it is well adapted as a scavenger and has taken well to being an urban raptor. Ruddy shelduck[04] refers to having pied colour. Bronze winged jacana[73] of both sexes is territorial in their nests, with the larger female observer having multiple mates and egg clutches. Bar headed goose[02] has a much smaller beak than the other large goose, giving it a more elegant appearance overall. Greylag goose[03] is believed to be the common ancestor of all domestic breeds. It makes for a beautiful sight when the whole flock decides to fly. Black tailed godwit[75] used to be popular game birds, due to their size and stimulating serial manoeuvres. Garganey[06] can be seen in groups, diving and feeding in a perpendicular sense. White tailed lapwing[71] is mostly silent unlike the red wattled lapwing[70]. Immature Bluethroat[170] doesn't have a visible blue throat and can easily disappear in dry undergrowth. Plain prinia[148] can be easily recognised by its tail. Black winged stilts[64] usually fly with their legs extended to the back. They usually occupy muddy shores and shallow waters, eating insects and invertebrates. White throated kingfisher[114] can have up to 4-5 of these in the same vicinity. Eurasian wigeon[09] can be found feeding in shallower parts of the wilder lakes and ponds. Citrine wagtail[139] appear as House sparrow[136] with tails almost as long as their bodies. They will fly in a bobbing pattern and will call as they fly. Oriental white eye[158] will keep hopping from branch to branch, tree to tree and it is almost impossible to find them sitting still. Grey breasted prinia[146] are much smaller than those seen around here. This is a restless bird, kept hopping from branch to branch in the half hour when we said this one. Prefer to stay near the edges of scrub forests. White wagtail[140] is similar in size to grey wagtail[138], this one has puffed its feathers early in the day. Finally found one sitting and catching insects. Brown shrike[123] is smaller in size than the long-tailed shrike. Black headed ibis[57] is feathered in juveniles but featherless in adults. They eat insects and invertebrates on both land and water. Pied bush chat[175] can also be a juvenile moulting into an adult. Red breasted flycatcher[171] nothing looks like its name but this is one. As they don't have the characteristic red breast. Black redstart[174] can make it look like a completely different bird, a winter migrant. Taiga flycatcher[172] have a habit of keeping their tail cocked. Like all flycatchers, it can be best spotted hopping branches near rotting vegetation and muck, where flies and smaller insects are frequent. Tickell's blue flycatcher[168] prefers dreier scrubby areas and forest edges. Ultramarine catcher[173] is a small stocky bird with deep blue black and a striking patch on throat, breast and belly in males. Painted stork[38] is easily recognisable for its size. Like other storks, they are known to not have a call and make sounds either by hitting the upper and lower beaks together (bill clattering) or making very soft calls. Common tailorbird[149] can be found in almost every park and garden, just as easily as it can be found in the wilderness. Common hawk cuckoo[30] to suggest like mimics (present in a number of cuckoos) help scare the host bird from its nest, so they can easily lay eggs. Glossy ibis[60] stay close to water bodies and feed near the shore. In breeding plume, the same become a glistening black. Ruff[76] is attractive wader which migrats to the park in winter near wetland. Rufous treepie[128] is a fun bird to observe. They can easily camouflage between leaves and their shadows. They have two prominent calls, a sweet song like call and a jarring laughter like call. Indian silverbill[134] is even smaller if we only measure its height. It can be identified by the shape of its beak. Long tailed shrike[126] is particularly good at mimicry, copying the calls of migratory birds, yelps of puppies, human whistling and even the shrieks of a frog to perfection! Grey wagtail[138] are winter visitors to most of India. Black bellied tern[84] prefers fresh water habitats and breeds on sandy islands. Red vented bulbul[155] is hard to tell if it is a different subspecies or needs to moult to an adult, but it is nice to notice a difference. Jacobin cuckoo[28] were found under tree cover, we were trying to find another bird, when we spotted this one from a gap in the canopy. Baya weaver[132] is usually seen in sizeable flocks, we spotted these two in isolation.

Stopped by, maybe, in passage to another place. Jungle babbler[160] near humans, where seeds and feed are abundant, almost wholly stick to being fed. Oriental honey buzzard[87] is migratory bird visits during winter. Short-toed eagle[90] is a snake eating bird. Intermediate egret[55] are similar in height to purple heron[53]. It appears taller though, being upright more often. Oriental magpie robin[167] has a pretty whistle like song, that is personally my favourite. Grey heron[52] can be seen active and upright by about 6:30-7am. As the day grows on, they sit with their neck retracted, usually at the far end of the park, away from people. Brahfminy starling[163] are seen in frequent trees and can be seen around most government colonies but interestingly they are not so frequent in woody region. White eared bulbul[154] is similar in size and appearance to the red vented bulbul[155], this one is browner in colour. Greater coucal[27] is closely related to the cuckoo but differs in that it raises its own young one. It is commonly referred to as the crow pheasant as well. Common moorhen[34] are alert and shy birds, so usually they flee before spotting them. Rose ringed parakeet[118] can fly with a noisy cackle and can often be heard before we spot them. They are one of the most wasteful eaters, they almost always drop more food than they eat. Purple heron[53] can fly between feeding spots and to find an undisturbed spot for the day. Are mostly inactive once they've fed and will sit slouched between tall grasses and marshes. Wire tailed swallow[152] has a tiny whistling call, a tinny version of rose ringed parakeet[118]. Flies in swift patterns with abrupt changes and is difficult to track. Bank myna[165] look similar to common myna[164] and are unaffected by humans, they do keep away from traffic though. Montagu's harrier[93] is winter visitor in grassland. Brown headed barbet[108] nest in February. Once they do lay eggs and later have young ones, they temporarily 'disappear', only to appear again once the nestlings have grown up. Indian black ibis[59] can be seen foraging for insects and fallen fruits. Don't easily fly away but are still best observed from a distance. Ashy prinia[147] dislikes people getting too close but always ends up observing before ignoring. Common koel[29] is slightly smaller than the adult koel. For Red wattled lapwing[70] we keep our distance and observe without disturbing their peace. The parents will chase us away, if we wait long enough for the observation. Northern lapwing[67] is spotted near the fields and farmlands, feeding on insects and worms of soil. Cattle egret[51], this one gets its name from its usual perch which is cattle. These aren't limited to areas with cattle and frequent grassy shrubbery and smaller tree tops while looking for insects. White breasted waterhen[32], may let people stay close. Indian spot billed duck[10] is a very alert bird that moves away as soon as it spot. Its twinkling eyes are really a beautiful sight. Yellow bittern[45] is hard to spot if you're already looking for it. Cinnamon bittern[46] usually ventures out during twilight hours and only during the daytime only when it is overcast. With its colour, it blends only too well with its surroundings and become hard to spot. Black bittern[47] is mainly venture in twilight hours or at night. Little swift[26] cannot perch on branches like most birds do. They tend to cling to rough surfaces, such as buildings and caves. Brown rock chat[178] is a friendly bird, doesn't mind humans until a distance but will flee at sudden movements. Laughing dove[21] is a beautiful bird with a pleasing coo. Little egret[56] is a part of special breeding plume and raises its wings as a mating dance. Great egret[54] is taller than the little egret[56] and similar to the grey heron[52] in size and habitat. Straited heron[49] is small widespread heron inhabits wetlands. Northern shoveler[07] beak is similar to a sieve and they strain the water taking in smaller insects. Little ringed plover[65] is small wader prefers freshy and bracky wetland. Kentish plover[66] is small sized migrant that may join flocks of other short birds in wetland. House crow[129] are intelligent, so much so that in some forms of Hinduism, they are recognised as reincarnations of our ancestors. Climate change is expected to have a major impact on biodiversity, including on the distribution of species. Global environmental commitments are in place but now must be realised. Moving forward we need to link biodiversity more clearly to people's livelihoods and well-being, to build constituencies for change, and to look after both key sites and the wider landscape. Birds are showing the way, and also helping us to keep track of progress.

CONCLUSION

Keoladeo National Park supports rich bird diversity due to its wetlands, forests, and grasslands. The study (Dec 2022–Mar 2023) recorded 179 bird species belonging to 18 orders and 62 families, including 102 resident and 77 migratory species, with Passeriformes being the most dominant order.

Several species show declining populations due to habitat loss, construction, and environmental disturbances. Threatened or declining birds include the Dalmatian Pelican, Northern Lapwing, Eurasian Curlew, Black-tailed Godwit, Black-necked Stork, Woolly-necked Stork, and Painted Stork. The Black-headed Ibis is also listed as Near Threatened due to population decline in some habitats. Overall, the study recorded 9 Near Threatened, 3 Vulnerable, 2 Endangered, and 165 Least Concern species.

Bird feeding habits influence habitat use: frugivores are mainly associated with forests, herbivores with wetlands, while insectivores and omnivores occur in both forests and other habitats. High bird diversity is supported by diverse habitats and food availability. However, human activities such as livestock grazing, deforestation, pollution, climate change, road construction, and resource extraction are major threats that reduce species richness and degrade habitats, making human disturbance a key factor affecting the ecological health of the park.

REFERENCES

1. Archibald, G. W. et al., (2003). A review of the three subspecies of Sarus Cranes *Grus antigone*. *Journal of Ecological Society*, *16,* 5–15.

2. Azman, N. M. et al., (2011). Avian diversity and feeding guilds in a secondary forest, an oil palm plantation and a paddy field in riparian areas of the Kerian River Basin, Perak, Malaysia. *Tropical Life Sciences Research*, *22*(2), 45–64.

3. Azeez, P. A. et al., (1992). The socio-economics of the villages around Keoladeo National Park, Bharatpur (Rajasthan), India. *International Journal of Ecology and Environmental Sciences*, *18,* 169–179.

4. Bhadouria, B. S. et al. (2019). Records of Avifauna of Wetlands around Keoladeo National Park, Bharatpur, India.

5. Bhadouria, B. S., et al. (2012). A survey of avifaunal diversity in wetlands around Keoladeo National Park, Bharatpur, Rajasthan, India. *Bird Populations*, *11,* 1–6.

6. Bhadouria, B. S. et al., (2012). Monitoring of organochlorine pesticides in and around Keoladeo national Park, Bharatpur, Rajasthan, India. *Environmental Monitoring and Assessment, 184,* 5295–5300,

7. Borad, C. K. et al. (2002). Breeding performance of Indian Sarus Crane *Grus antigone antigone* in the paddy crop agroecosystem. *Biodiversity & Conservation*, *11*, 795–805.

8. Chauhan, M. & Gopal, B. (2001a). Biodiversity and management of Keoladeo National Park (India): a wetland of international importance. *Biodiversity in wetlands: assessment, function and conservation, 2*, 217–256.

9. Chauhan, M. and Gopal, B. (2001b) 'Biodiversity and management of Keoladeo National Park (India): a wetland of international importance', *Biodiversity in wetlands: assessment, function and conservation*, 2, pp. 217–256.

10. Choudhary, A., et al., (2021). Study of habitat quality assessment using geospatial techniques in Keoladeo National Park, India. *Environmental Science and Pollution Research*, *28,* 14105–14114.

11. Das, J. & Parida, S.P. (n.d.). A studies on wintering wetland avifauna of Sundargarh forest division, Sundargarh, India. *Research Journal of Recent Sciences.*

12. Dookia, S. (2015). Ecology and Behaviour of Indian Peafowl (Pavo cristatus) in Keoladeo National Park, Bharatpur, Rajasthan, India. *International Journal of Fauna and Biological Studies*, *2*(4), 99–105.

13. Kulshreshtha, S. et al., (2013). The Ramsar sites of Rajasthan: ecology and conservation of Sambhar Salt Lake, Jaipur and Keoladeo National Park, Bharatpur. *Faunal Heritage of Rajasthan, India: Conservation and Management of Vertebrates*, 173–219.

14. Lawania, K.K. & Trigunayat, M.M. (2015). A comparative study of the spider (Araneae) fauna in Keoladeo National Park (KNP), Nahargarh Wildlife Sanctuary (NWS) and Sur-sarovar Bird Sanctuary (SBS), India. *International Journal on Agricultural Sciences*, *6*(1), 141–146.

15. Middleton, B. A. (1989). *Succession and goose herbivory in monsoonal wetlands of the Keoladeo National Park, Bharatpur, India* [Doctoral dissertation, Iowa State University].

16. Muralidharan, S. (1993). Aldrin poisoning of Sarus cranes (Grus antigone) and a few granivorous birds in Keoladeo National Park, Bharatpur, India. *Ecotoxicology, 2,* 196–202.

17. Neeraj, S. et al., (2016). Effect of changing hydrological parameter on declining of species and population of birds in Keoladeo National Park Bharatpur wetland. *International Journal of Pharmacology and Biological Sciences, 10*(1), 19.

18. Pain, D.J. et al. (2004). Mortality of globally threatened Sarus Cranes Grus antigon from monocrotophos poisoning in India. *Science of the Total Environment, 326*(1–3), 55–61.

19. Patra, A. et al., (2017). Modeling and simulation of a wetland park: An application to Keoladeo National Park, India. *Mathematics and Computers in Simulation, 134,* 54–78.

20. Puneet, P. (n.d.). Sarus Crane population fluctuation at various wetlands at Bharatpur in Rajasthan State of India.

21. Reddy, M.S. & Char, N.V.V. (2006). Management of lakes in India. *Lakes & Reservoirs: Research & Management, 11*(4), 227–237.

22. Registro, G. (n.d.). Integrated Publishing Toolkit (IPT).

23. Saluja, R. & Garg, J.K. (2017). Trophic state assessment of Bhindawas Lake, Haryana, India. *Environmental monitoring and assessment, 189,* 1–15.

24. *Second Annual Research Seminar–KNP.*

25. Sharma, K.K. et al. (n.d.). International Journal of Environment & Animal Conservation.

26. Sharma, N. et al., (2015). Effects of hydrological changes on the biodiversity at Keoladeo National Park and their impact on Ecotourism. *International Journal of Civil, Structural, Environmental and Infrastructure Engineering Research and Development (IJCSEIERD).*

27. Singh, A. *et al.* (2017) 'An updated account of mammal species and population status of ungulates in Keoladeo National Park, Bharatpur, Rajasthan', *Current Science*, pp. 103–111.

28. Srivastava, H.S. et al. (2007). Potential applications of multi-parametric synthetic aperture radar (SAR) data in wetland inventory: a case study of Keoladeo National Park (A World Heritage and Ramsar site), Bharatpur, India. *Proceedings of Taal*, 1862–1879.

29. Srivastava, H.S. et al. (2009). Explored and demonstrated potential applications of multiparametric synthetic aperture radar in wetland studies in context of Keoladeo National Park, Bharatpur, India. In *Proc.*

30. *The IUCN Red List of Threatened Species* (no date) *IUCN Red List of Threatened Species*. Available at: https://www.iucnredlist.org/en (Accessed: 28 May 2023).

31. Tiwari, S.K. & Upadhyay, R.K. (2017). Conservation of degraded wetland system of Keoladeo National Park, Bharatpur, India. *Ecological Complexity*, *32*, 74–89.

32. Upadhyay, R. K. et al., (2014). Modeling wetland systems of Keoladeo National Park (KNP), India: the role of space. *Wetlands ecology and management*, *22*, 605–624.

33. Urfi, A. J. et al. (2005). Counting birds in India: Methodologies and trends. *Current Science*, 1997–2003.

Chapter - 8

EFFECT OF DIFFERENT HEPATOCELLULAR CARCINOMA INDUCING AGENTS IN ANIMAL MODELS

Dr. Nighat Gull[1] and Prof. Arif Ahmad[2*]

[1,2]Department of Zoology, School of Science,

MANUU Gachibowli, Hyderabad,
Telangana 500032, India

Corresponding email: arifamu@gmail.com

ABSTRACT

Liver cancer continues to pose a significant global health challenge, with projections indicating an incidence exceeding 1 million cases by 2025. Hepatocellular carcinoma is the predominant form of liver cancer, accounting for approximately 90% of cases. Chronic hepatitis and cirrhosis are the primary pre-neoplastic states in a substantial majority of hepatocellular carcinoma instances and may be linked to other causative agents, including environmental chemical carcinogens. Creating suitable hepatocellular carcinoma animal models is crucial to advance both fundamental and translational research, especially models that can precisely mimic specific human disease circumstances. Current animal models are categorized as chemically-induced, genetically-engineered, and diet-induced models. This chapter revives use of different hepatocellular carcinoma-inducing agents in animal models, focusing on their processes, effects, and significance to human pathology. This study investigates significant chemical carcinogens, including Diethylnitrosamine (DEN), carbon tetrachloride (CCl_4), and aflatoxins, regarding their ability to cause liver damage, fibrosis, and tumor formation, thereby providing insights into the chemical origins of hepatocellular carcinoma. The inquiry into the function of viral infections, specifically hepatitis B (HBV) and hepatitis C (HCV) use animal models that replicate the chronic inflammation, fibrosis, and oncogenesis linked to these viruses. Moreover, dietary variables and environmental exposures, including high-fat diets, alcohol use, and metabolic disorders, are examined as contributors to the rising prevalence of hepatocellular carcinoma.

Keywords: *Hepatocellular Carcinoma, Animal models, Diethylnitrosamine (DEN), carbon tetrachloride (CCl_4), and aflatoxins, hepatitis B (HBV), hepatitis C (HCV)*

1. INTRODUCTION

Liver cancer ranks as the sixth most prevalent malignancy and is the third leading cause of cancer-related mortality globally (Tan et al., 2021). Hepatocellular carcinoma is the primary histological subtype of liver cancer, representing 80–85% of cases (Kulki & Serag, 2019). An estimated 900,000 individuals worldwide were diagnosed with liver cancer in 2020, and approximately 830,000 died from the disease. Eastern Asia, North Africa, and Southeast Asia recorded the highest rates of liver cancer incidence and mortality (Devarbhavi et al., 2023). In 15 countries, particularly in regions such as Southeast Asia and sub-Saharan Africa, hepatocellular carcinoma is the leading cause of cancer-related mortality. The global burden of liver cancer is projected to increase substantially over the coming decades, with forecasts indicating a significant rise in both new cases and deaths if current trends persist (Rumgay et al., 2022).

In recent years, drug-targeted therapies have increasingly become a focal point of research, with hepatocellular carcinoma being influenced by numerous pathogenic factors and multiple pathways involved in its development and progression. (Li et al., 2023). Preclinical animal models are thought to be useful and trustworthy resources for studying the pathophysiology, molecular mechanisms underlying hepatocarcinogenesis, identifying potential therapeutic targets, performing drug screening, and assisting in the assessment of the effectiveness of new treatments for hepatocellular carcinoma (Li et al.,2021; Brown et al.,2018). The liver is made up of tiny structures known as lobules, which are thought to be both structural and functional components of the organ. In the event of an infection or damage, liver function is impaired. Drug-induced liver damage (DILI), Aflatoxin B1-contaminated food consumption, autoimmune hepatitis, nonalcoholic steatohepatitis (NASH), alcoholic cirrhosis or viral hepatitis, and obesity/type 2 diabetes (Gingold et al., 2018). Furthermore, the incidence of liver cancer may also be correlated with age, gender, and geographic location (Qiu et al., 2019; Sayiner et al., 2019) and other conditions might fit this description.

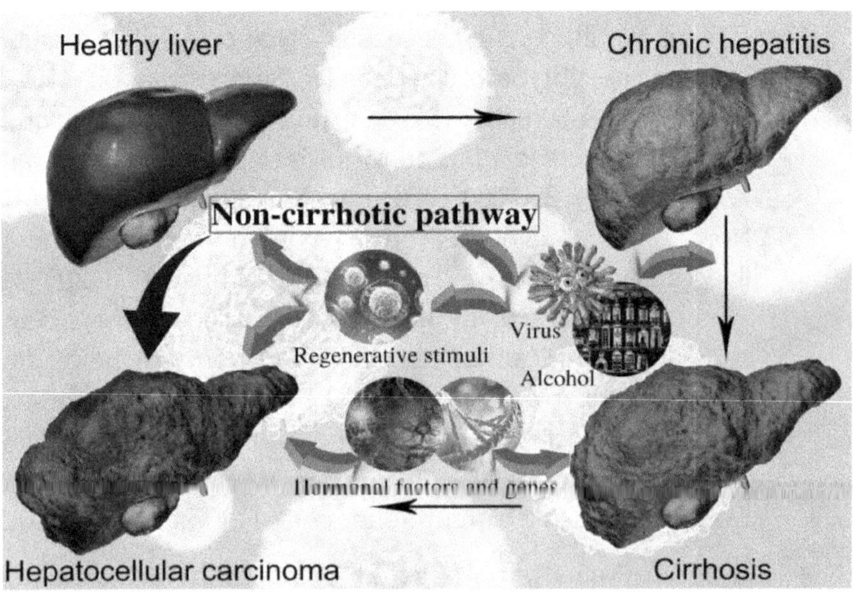

Figure 1. Schematic representation of liver cancer development and progression (Source Nighat et al., 2022).

2. ANIMAL MODELS FOR LIVER CANCER STUDIES

A key connection between cell lines and clinical research is made by animal investigations due to similarities in anatomy and physiology of humans. Inaccurate model selection results in inaccurate conclusions (Santosh et al., 2017). At present, frequently used hepatocellular carcinoma animal models are zebra fish, mice, rats, rabbits, and woodchucks because they show genetic similarities and share biological characteristics with humans, used to study quantitative imaging techniques and intervention treatments (Lee et al.,2021; Keller et al.,2020; Pritchard et al.,2020; Finnberg et al.,2004). Animal models can categorize in:

1. Genetically engineered models
2. Dietary induced models
3. Chemically-induced models

2.1. Genetically engineered models (GEM)

For deciphering the intricacies of cancer biology and testing out new treatment approaches, genetically modified models are essential resources (Henderson et al., 2017). GEM of hepatocarcinogenesis, which emphasize single or multiple mutations, have been extensively utilized, Furthermore, studies have presented models that incorporate modifications to the tumor microenvironment or growth factors. Both conditional GEMs and knockout models, which are employed in loss-of-function investigations, have been made possible by gene targeting technologies (Heindryckx & Vlierberghe (2009). These models make use of site-specific recombinase systems to precisely regulate gene expression in a limited amount of time and space (Liu et al., 2017). The Clustered Regularly Interspaced Short Palindromic Repeats-associated protein 9 (CRISPR/Cas 9) a, revolutionary development in gene editing technology, allows precise alterations to DNA sequences within living things (Callegari et al., 2012; Katz al., 2012). Advanced techniques, including as tetracycline-controlled systems, microRNA distribution, and Cre-LoxP recombination, offer fine-grained control over gene expression and the targeted insertion of genetic changes related to liver. Generation of transgenic mice models harbouring distinct segments of the hepatitis viral genome has evolved into a pivotal research tool for elucidating the pathogenic effects of human hepatitis virus on hepatocellular carcinoma. GEM have multifaceted applications in tumor biology research investigating tumorigenesis, disease development and metastatic spread (Li et al., 2023).

2.2. Dietary induced models

Intake of high diet fat (HFD) is undoubtedly linked to a number of detrimental effects on liver function, including a considerably elevated risk of hepatocellular carcinoma (Nevzorova et al., 2020). HFD-induced accumulation of hepatic lipids initiates an inflammatory cascade concurrently triggering endoplasmic reticulum stress and lipotoxicity in hepatocytes, resulting in hepatocyte damage and ultimately contributing to the spontaneous development of hepatocellular carcinoma (Peiseler et al., 2022). Moreover, HFD may facilitate hepatocarcinogenesis by promoting the upsurge of acylcarnitine and exacerbating certain glucose-mediated metabolic changes (Broadfield et al., 2021; Fujiwara et al., 2018). Numerous epidemiological studies have provided evidence of a correlation between dietary cholesterol consumption and the increased risk and severity of non-alcoholic fatty liver disease. (Zhang et al., 2021). Elevated cholesterol levels led to significant increases in serum leptin, interleukin-6, liver weight, and insulin resistance, this resulted in a condition that closely mimics the clinical manifestations of NASH observed in individuals with metabolic syndrome (Henkel et al., 2017). With an extended induction period and a diet rich in fat, fructose, and cholesterol, combined with a low weekly dose of intraperitoneal CCL^4 as a promoter, a murine NASH model shows rapid advancement of advanced fibrosis by 12 weeks and hepatocellular carcinoma by 24 weeks. This model reflects the histological, immunological, and transcriptomic characteristics of human NASH (Tsuchida et al., 2018).

Choline-deficient (CCD) diet has been widely utilized in animal investigations of non-alcoholic fatty liver disease (NAFLD) (Kesarwala et al., 2016). CCD also imitates the pathogenic mechanism underlying the initiation of non-alcoholic steatohepatitis (NASH) and its progression to more severe stages of hepatocellular carcinoma (Knight et al., 2000). Combining CDD with other carcinogenic agents in hepatocellular carcinoma animal model development may reduce induction time and enhance success rates. A significant increase in alpha1-fetoprotein levels was seen in the liver and plasma of rats administered CDD and DEN, expediting tumor development (Delima et al., 2008).

2.3. Chemically-induced models

Liver plays a vital function in detoxifying harmful chemicals exposed to humans (Thompson 2021). Numerous recognized chemically induced agents are known to generate liver damage and are thus employed as promoters in hepatocellular carcinoma models. These substances primarily cause hepatocyte apoptosis, steatosis, or cholestasis, which later results in inflammation and fibrosis (Liu et al.,2015; Wang et al., 2012).

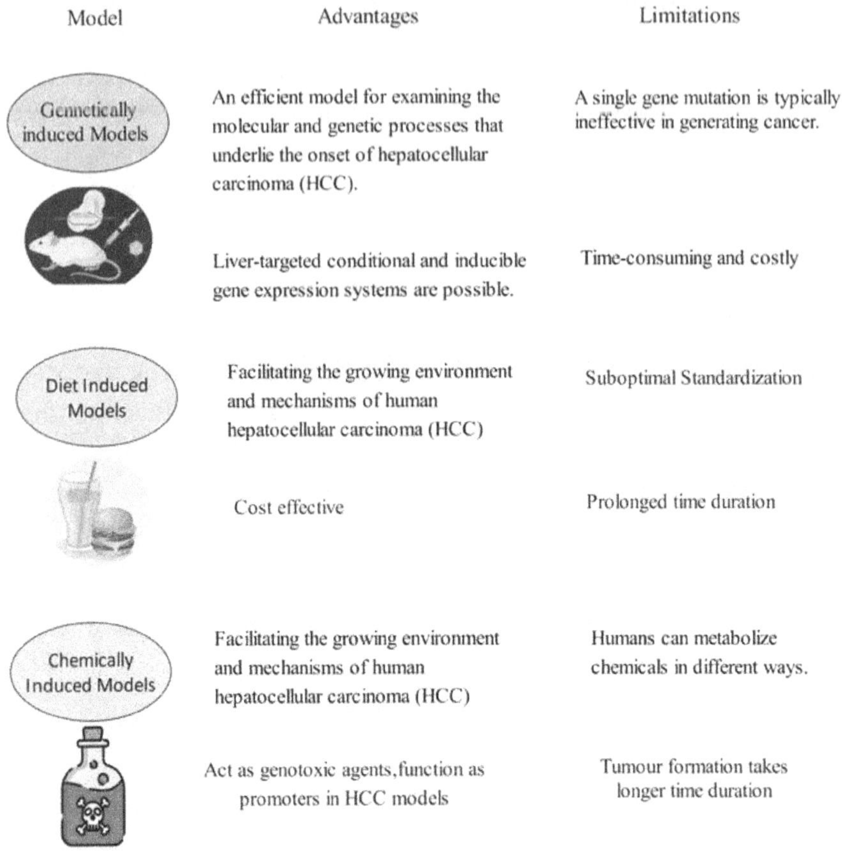

Figure 2. Advantages and Limitations Genetically induced, Diet induced and Chemically induced Models

3. CHEMICAL CARCINOGENS AS LIVER CANCER-INDUCING AGENTS

3.1. Diethylnitrosamine (DEN)

Diethylnitrosamine (DEN) has been detected in a diverse array of agricultural products, encompassing preservatively-treated seafood, processed meat items, dairy products, and various oil-based commodities (Park et al., 2015). DEN during its metabolic process, produces free radicals and disrupts the equilibrium of antioxidants, leading to oxidative stress (Latief et al., 2016; Mukherjee and Ahmad 2018). The process of free radical production requires the bioactivation of DEN by a specific group of hepatic cytochrome P450s, as described by Archer in 1989. As a result, O4- and O6-ethyldeoxythymidine, and O6-ethyldeoxyguanosine are produced, which are ethyl radical metabolites (CH3CH2+).Typically, DEN is employed to induce liver tumours in rodents, as a model, to understand the fundamental mechanism of the disease, but can also develop hematopoietic, respiratory skin, and gastrointestinal tumours (Husain et al., 2018; Latief et al., 2016; Ahmad and Ahmad, 2014; Bansal et al., 2005). Earlier research on rodents had shown that hepatocellular carcinogenesis can be induced by a single intraperitoneal dosage of DEN (200 mg kg-1 body weight) (Singh et al., 2018; Saraswati et al., 2013) In mice older than two weeks, the administration of DEN necessitates a tumor promoter to augment hepatocarcinogenesis. This may be achieved by utilizing the following promoters: 2-AAF, phenobarbital (PB), N-nitrosomorpholine (NMOR), and carbon tetrachloride (CCL^4). DEN-induced hepatocarcinogenesis depends on the dosage, sex, age, mouse strain and duration of exposure.

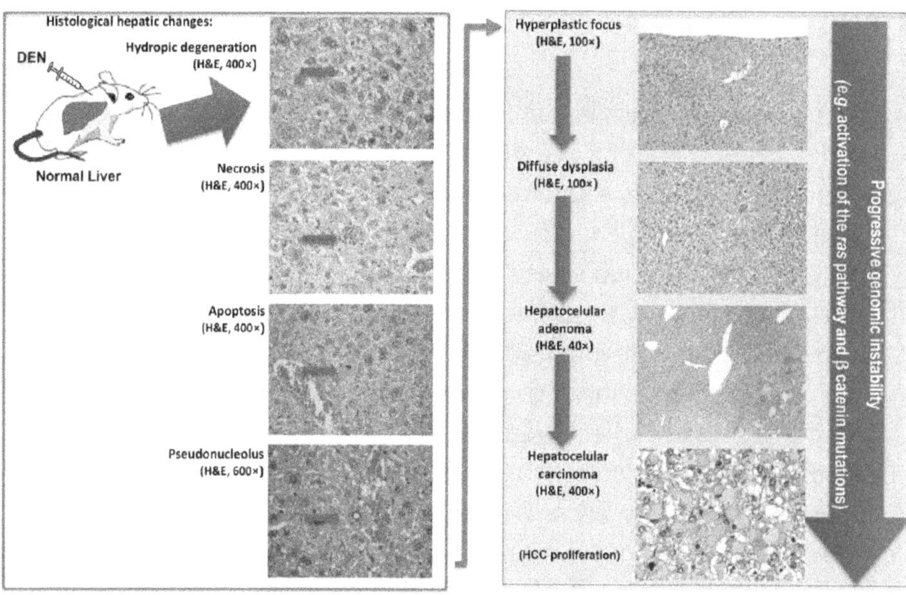

Figure 3. Histopathological changes observed in DEN treated animals (Adapted from Santos et al., 2017)

3.2. Carbon tetrachloride (CCL⁴)

Carbon tetrachloride (CCL^4) is a potent hepatotoxin (Bala et al., 2012) used as a promoter and accelerates the progression of liver cancer in a shorter duration of time (Uehara et al., 2013). The principal pathways of possible human exposure to CCL^4 include inhalation, ingestion, and skin contact via air and water. Usually, 0.5 to 2 mL/kg body weight of CCL^4 (diluted in paraffin oil) is given or maize oil) via intraperitoneal injection twice a week, resulting in a strong and repeatable fibrosis within four to six weeks. Within a few weeks, CCL^4-induced liver fibrosis can be fully recovered (Heindryckx & Vlierberghe (2009). Using CCL^4, several rat models have been shown to develop both liver cirrhosis and hepatocellular carcinoma (Elaidy et al., 2017). The toxicity of CCL^4 primarily relies on the breaking of the carbon-chlorine bond, resulting in the formation of trichloromethyl radicals. The formation of a trichloromethyl substitution free radical is produced, when a free radical reacts fast with oxygen. This process is associated with an increase in hepatotoxicity and the subsequent generation of liver enzymes, resulting in severe liver damage (Basu et al., 2011).

3.3. Aflatoxins

Hepatocellular Carcinoma carcinogenesis is associated with exposure to various environmental chemicals, among which aflatoxin is the most prominent. Aflatoxins, a class of mycotoxins predominantly produced by the fungi "*Aspergillus flavus*" and "*Aspergillus parasiticus*" pose a major risk to global food safety, especially in warm and humid climates that favor fungal growth during crop cultivation (Wu & Santella 2012). Aflatoxin B1 is a prevalent environmental contaminant of dietary staple crops, demonstrating potent carcinogenic, mutagenic, and immunosuppressive effects (Cao et al., 2022). Multiple variables have been recognized as synergistically interacting with both HBV and HCV infections. Numerous recent research offers rising epidemiological evidence linking AFB1 exposure to advanced liver disease in persons infected with HCV (Jin et al., 2023). Aflatoxin exposure and chronic HBV infection are more prevalent among rural inhabitants than urban residents in resource-limited areas. The correlation between aflatoxin exposure and the onset of hepatocellular carcinoma is most pronounced in Sub-Saharan Africa (Kew, 2003).

4. BIOLOGICAL AGENTS INDUCING LIVER CARCINOGENESIS

4.1. Hepatitis B virus (HBV)

Chronic infection with the HBV is identified as a major risk factor in the etiology of hepatocellular carcinoma, with an estimated prevalence of approximately 50% of worldwide cases. Hepatitis B virus (HBV) elevates the risk of hepatocellular carcinoma (HCC) even in the absence of cirrhosis, although the majority of patients with HBV-induced HCC present with cirrhosis. However, risk factors differ substantially across geographical regions, with notable variations observed in East Asia compared to Europe (Abdelhamed and Kassas 2024). Globally, an estimated 250 million individuals are subjects of chronic HBV exposure, rendering infants a considerable high-risk demographic. In regions identified as high-risk, the prevalence of HBV infection approximates 8% of the population, whereas amongst hepatocellular carcinoma patients, a notably high proportion, nearly 80%, exhibit the HBV surface antigen (Gull et al., 2022; El-Serag et al., 2012). Incidence of cirrhosis and hepatocellular carcinoma in untreated chronic HBV infection is substantial. Approximately 8-20% of patients develop cirrhosis within five years, with a concurrently elevated 2-5% annual risk of hepatocellular carcinoma in the presence of cirrhosis. Notably, HBV-related hepatocellular carcinoma is a leading cause of mortality, accounting for nearly 900,000 fatalities annually (Singal et al., 2023). Estimates suggest that an additional 5.3 million children are anticipated to develop chronic HBV between 2020 and 2030 as a direct consequence of the COVID-19 pandemic, with approximately 1 million projected fatalities globally. Prior to the implementation of vaccination programmes, HBV infection prevalence among children beneath the age of five was approximately 5%, decreasing significantly to around 1% by 2019 (Gull et al., 2022; Galibraith et al.,2015).

4.2. Hepatitis C virus (HCV)

HCV is a hepatocellular-associated, positive-sense single-stranded RNA virus that contributes to the development of in conjunction with other etiological factors. Incidence rates of HCV-related hepatocellular carcinoma exhibit the highest mortality rates among 100,000 inhabitants within the United States (Rafati et al., 2022). Epidemiological data indicate a significant association between HCV etiology and hepatocellular carcinoma, with prevalence rates differing between Asian and Caucasian populations, exhibiting notable disparities in the frequency of HCV-related hepatocellular carcinoma. Furthermore, an upward trend in the incidence of acute HCV cases among young individuals was observed during the 2009–2018 period, with individuals aged 20–29 years being disproportionately affected. According to a Center for Disease Control and Prevention (CDC) analysis, the incidence of cirrhosis is predominantly observed in patients with hepatocellular carcinoma associated with HCV, representing an approximate frequency of 80-90% (Serfaty, 2008). Furthermore, meta-analysis studies have revealed that individuals infected with HCV genotype 3 exhibit a 50% elevated rate of fibrosis compared to those harbouring alternate genotypes (Yang & Robert 2010).

4.3. Hepatitis D Virus (HDV)

Hepatitis D virus is an RNA-based pathogen that relies on the co-expression of HBV surface antigens for its replicative and infectious processes to occur. Global prevalence of hepatitis delta virus infection ranges approximately between 20 to 40 million individuals. Individuals infected with HDV exhibit a more accelerated progression of liver disease, marked by enhanced fibrosis and an elevated risk of cirrhosis relative to those infected with HBV in isolation. Research has shown that concurrent infection with HBV and hepatitis HDV increases the probability of liver cancer incidence, particularly hepatocellular carcinoma, in comparison to infection with HBV alone (Sausen et al., 2022). The incidence of hepatocellular carcinoma was considerably elevated among individuals afflicted with acute HDV infection (relative risk: 6.1, 95% confidence interval: 2.8-11.7) as well as persistent HDV infection (relative risk: 3.9,

95% confidence interval: 1.6-7.2) when compared to those harbouring HBV infection solely (Puigvehi et al., 2019).

CONCLUSION

This chapter elucidates the disparate consequences of hepatocellular carcinoma-inducing substances on animal models, thereby underscoring the uniqueness of carcinogenic pathways mediated by distinct chemical agents, including diethylnitrosamine, aflatoxins, and carbon tetrachloride, and the diverse tumorigenic potencies exhibited by these compounds. The heterogeneity in tumor progression and histopathological characteristics implies that each model offers distinct perspectives on the etiology of hepatocellular carcinoma, thereby providing valuable platforms for evaluating potential therapeutic interventions.

REFERENCES

1. Tan, E. Y. et al., (2024). Liver cancer in 2021: global burden of disease study. *Journal of Hepatology*.
2. Kulik, L. & El-Serag, H. B. (2019). Epidemiology and management of Hepatocellular Carcinoma. *Gastroenterology*, *156*(2), 477-491.
3. Devarbhavi, H. et al., (2023). Global burden of liver disease: 2023 update. *Journal of hepatology*, *79*(2), 516-537.
4. Rumgay, H. et al., (2022). Global burden of primary liver cancer in 2020 and predictions to 2040. *Journal of hepatology*, *77*(6), 1598-1606.
5. Li, J. et al., (2023). Advances in experimental animal models of Hepatocellular Carcinoma. *Cancer Medicine*, *12*(14), 15261-15276.
6. Brown, Z. J. et al., (2018). Mouse models of Hepatocellular Carcinoma: an overview and highlights for immunotherapy research. *Nature reviews Gastroenterology & hepatology*, *15*(9), 536-554.
7. Blidisel, A. et al., (2021). Experimental models of hepatocellular carcinoma-a preclinical perspective. *Cancers*, *13*(15), 3651.

8. Li, Z. et al., (2021). Application of animal models in cancer research: Recent progress and future prospects. *Cancer management and research*, 2455-2475.

9. Gingold, J. A. et al., (2018). Genomic profiling and metabolic homeostasis in primary liver cancers. *Trends in molecular medicine, 24*(4), 395-411.

10. Qiu, Z. et al., (2019). A pharmacogenomic landscape in human liver cancers. *Cancer Cell, 36*(2), 179-193.

11. Sayiner, M. et al., (2019). Disease burden of Hepatocellular Carcinoma: a global perspective. *Digestive diseases and sciences, 64,* 910-917.

12. Santos, N. P. et al., (2017). Animal models as a tool in Hepatocellular Carcinoma research: A Review. *Tumor Biology, 39*(3), 1010428317695923.

13. Finnberg, N. et al., (2004). Heterozygous p53-deficient (+/–) mice develop fewer p53-negative preneoplastic focal liver lesions in response to treatment with diethylnitrosamine than do wild-type (+/+) mice. *Cancer letters, 207*(2), 149-155.

14. Keller, S. et al., (2020). Quantitative MRI for assessment of treatment outcomes in a rabbit VX2 hepatic tumor model. *Journal of Magnetic Resonance Imaging, 52*(3), 668-685.

15. Pritchard, W. F. et al (2020). Transarterial chemoembolization in a woodchuck model of Hepatocellular Carcinoma. *Journal of Vascular and Interventional Radiology, 31*(5), 812-819.

16. Lee, A. Q. et al., (2021). Inducible liver cancer models in transgenic zebrafish to investigate cancer biology. *Cancers, 13*(20), 5148.

17. Henderson, J. M. et al., (2017). Hepatocellular Carcinoma: Mouse models and the potential roles of proteases. *Cancer letters, 387,* 106-113.

18. Heindryckx, F. et al., (2009). Experimental mouse models for Hepatocellular Carcinoma research. *International journal of experimental pathology, 90*(4), 367-386.

19. Liu, Y. et al., (2017). CRISPR/Cas9-mediated p53 and Pten dual mutation accelerates hepatocarcinogenesis in adult hepatitis B virus transgenic mice. *Scientific reports*, *7*(1), 2796.

20. Katz, S. F. et al., (2012). Disruption of Trp53 in livers of mice induces formation of carcinomas with bilineal differentiation. *Gastroenterology*, *142*(5), 1229-1239.

21. Callegari, E. et al., (2012). Liver tumorigenicity promoted by microRNA-221 in a mouse transgenic model. *Hepatology*, *56*(3), 1025-1033.

22. Nevzorova, Y. A. et al., (2020). Animal models for liver disease–a practical approach for translational research. *Journal of hepatology*, *73*(2), 423-440.

23. Peiseler, M. et al., (2022). Immune mechanisms linking metabolic injury to inflammation and fibrosis in fatty liver disease–novel insights into cellular communication circuits. *Journal of hepatology*, *77*(4), 1136-1160.

24. Broadfield, L. A. et al., (2021). Fat induces glucose metabolism in no transformed liver cells and promotes liver tumorigenesis. *Cancer research*, *81*(8), 1988-2001.

25. Fujiwara, N. et al., (2018). CPT2 downregulation adapts hepatocellular carcinoma to lipid-rich environment and promotes carcinogenesis via acylcarnitine accumulation in obesity. *Gut*, *67*(8), 1493-1504.

26. Zhang, X. et al., (2021). Dietary cholesterol drives fatty liver-associated liver cancer by modulating gut microbiota and metabolites. *Gut*, *70*(4), 761-774.

27. Henkel, J. et al., (2017). Induction of steatohepatitis (NASH) with insulin resistance in wild-type B6 mice by a western-type diet containing soybean oil and cholesterol. *Molecular Medicine*, *23*, 70-82.

28. Tsuchida et al., (2018). A simple diet-and chemical-induced murine NASH model with rapid progression of steatohepatitis, fibrosis and liver cancer. *Journal of hepatology*, *69*(2), 385-395.

29. Ma, C. et al., (2016). NAFLD causes selective CD4+ T lymphocyte loss and promotes hepatocarcinogenesis. *Nature, 531*(7593), 253-257.

30. Knight, B. et al., (2000). Impaired preneoplastic changes and liver tumor formation in tumor necrosis factor receptor type 1 knockout mice. *Journal of Experimental Medicine, 192*(12), 1809-1818.

31. De Lima, V. M. et al., (2008). A rodent model of NASH with cirrhosis, oval cell proliferation and Hepatocellular Carcinoma. *Journal of hepatology, 49*(6), 1055-1061.

32. Thompson, B. C. (2021). *Image Based Dosimetry Post Transarterial Radioembolization* (Doctoral dissertation, Wake Forest University).

33. Wang, L. et al., (2012). Hepatic vascular endothelial growth factor regulates recruitment of rat liver sinusoidal endothelial cell progenitor cells. *Gastroenterology, 143*(6), 1555-1563.

34. Liu, C. et al., (2015). Combination of sorafenib and gadolinium chloride (GdCl3) attenuates dimethylnitrosamine (DMN)-induced liver fibrosis in rats. *BMC gastroenterology, 15*, 1-11.

35. U.S. Department of Health and Human Services, Public Health Service, National Toxicology Program. *Report on Carcinogens*. 12th ed. (ed Lunn RM). Research Triangle

36. Park, M. J. et al., (2012). Small Hepatocellular Carcinomas: improved sensitivity by combining gadoxetic acid–enhanced and diffusion-weighted MR imaging patterns. *Radiology*, 264(3), 761-770.

37. Latief, U. et al., (2016). Hepatoprotective efficacy of gallic acid during nitrosodiethylamine-induced liver inflammation in Wistar rats. *The Journal of Basic & Applied Zoology*, 76, pp.31-41.

38. Mukherjee, D. & Ahmad, R. (2018). Resveratrol attenuates Nitrosodiethylamine-induced liver injury in anti-inflammatory manner via regulating cyclooxygenase-2. *Journal of Food Biochemistry,* 425), e12594.

39. Archer, M.C. (1989). Mechanisms of action of N-nitroso compounds. *Cancer surveys*, 8(2), pp.241-250.

40. Bansal, A. K. et al., (2005). Protective role of Vitamin E pre-treatment on N-nitrosodiethylamine induced oxidative stress in rat liver. *Chemico-biological interactions, 156*(2-3), pp.101-111.

41. Ahmad,A.&Ahmad,R.,(2014).Resveratrolmitigatestructuralchanges and hepatic stellate cell activation in N′-nitrosodimethylamine-induced liver fibrosis via restraining oxidative damage. *Chemico-biological interact tions, 221*, pp.1-12.

42. Husain, H. et al., (2018). Pomegranate action in curbing the incidence of liver injury triggered by Diethylnitrosamine by declining oxidative stress via Nrf2 and NFκB regulation. *Scientific reports,* 8(1), p.8606.

43. Saraswati, S. et al., (2013). Anticarcinogenic effect of brucine in diethylnitrosamine initiated and phenobarbital-promoted hepatocarcinogenesis in rats. *Chemico-biological interactions,* 206(2), 214-221.

44. Singh, D. et al., (2018). Attenuation of diethylnitrosamine (DEN)–Induced hepatic cancer in experimental model of Wistar rats by Carissa carandas embedded silver nanoparticles. *Biomedicine & Pharmacotherapy*, 108, 757-765.

45. Bala, A. et al., (2012). Carbon tetrachloride: a hepatotoxin causes oxidative stress in murine peritoneal macrophage and peripheral blood lymphocyte cells. *Immunopharmacology and Immunotoxicology*, 34(1), 157-162.

46. Takeki, et al., (2013). Molecular mechanisms of fibrosis-associated promotion of liver carcinogenesis." *Toxicological sciences* 132.1 (2013): 53-63.

47. Elaidy, S. M. et al., (2017). Evaluation of the therapeutic effects of Polyvinylpyrrolidone-capped silver nanoparticles on the Diethylnitrosamine/carbon tetrachloride-induced Hepatocellular Carcinoma in rats. *Egyptian Journal of Basic and Clinical Pharmacology,* 7(2), 9-24.

48. Basu, S. (2011). Carbon tetrachloride-induced hepatotoxicity: a classic model of lipid peroxidation and oxidative stress. *Studies on experimental models*, 467-480.

49. Wu, H. C., & Santella, R. (2012). The role of aflatoxins in Hepatocellular Carcinoma. *Hepatitis monthly*, *12*(10) *Hepatocellular Carcinoma*, e7238.

50. Cao, W. et al., (2022). Aflatoxin B1: Metabolism, toxicology, and its involvement in oxidative stress and cancer development. *Toxicology Mechanisms and Methods*, *32*(6), 395-419.

51. Jin, J. et al., (2023). Synergism in actions of HBV with aflatoxin in cancer development. *Toxicology*, *499*, 153652.

52. Kew, M. C. (2003). Synergistic interaction between aflatoxin B1 and hepatitis B virus in hepatocarcinogenesis. *Liver international*, *23*(6), 405-409.

53. Abdelhamed, W. & El-Kassas, M. (2024). Hepatitis B virus as a risk factor for Hepatocellular Carcinoma: there is still much work to do. *Liver Research*.

54. El-Serag, H. B. (2012). Epidemiology of viral hepatitis and Hepatocellular Carcinoma. *Gastroenterology*, 142(6), 1264-1273.

55. Nighat, G. et al., (2023). Recent advances in anticancer activity of novel plant extracts and compounds from Curcuma longa in Hepatocellular Carcinoma. *Journal of Gastrointestinal Cancer* 54.2: 368-390.

56. Singal, A. G. et al., (2023). Global trends in Hepatocellular Carcinoma epidemiology: implications for screening, prevention and therapy. *Nature reviews Clinical oncology*, *20*(12), 864-884.

57. Galbraith, J. W. et al., (2015). Unrecognized chronic hepatitis C virus infection among baby boomers in the emergency department. *Hepatology*, 61(3), 776-782.

58. Rafati, A. et al., (2022). Stem cells as an ideal carrier for gene therapy: A new approach to the treatment of hepatitis C virus. *Transplant Immunology*, *75*, 101721.

59. Serfaty, L. & Lemoine, M. (2008). Defnition and natural history of metabolic steatosis: clinical aspects of NAFLD, NASH and cirrhosis. *Diabetes Metab*.

60. Yang, J. D. & Roberts, L.R. (2010). Hepatocellular Carcinoma: a global view. *Nature Reviews Gastroenterology & Hepatology.*

61. Sausen, D. G. et al., (2022). Hepatitis B and Hepatitis D Viruses: A Comprehensive update with an immunological focus. *International Journal of Molecular Sciences, 23*(24), 15973.

62. Puigvehí, M. et al., (2019). The oncogenic role of hepatitis delta virus in Hepatocellular Carcinoma. *JHEP Reports*, 1(2), 120-130.

Chapter - 9

ENVIRONMENTAL POLLUTION: HEAVY METAL IONS CONTAMINATION

Dr. Preeti Chaudhary[1] and Dr. Dharmendra Singh[2]

[1] Department of Chemistry, Raja Mahendra Pratap Singh University Aligarh, U. P.

[2] Department of Chemistry, Dharma Samaj College Aligarh, U. P.

Corresponding email: dhsiitd@gmail.com

ABSTRACT

In order to improve water quality and lessen the effects on human and ecosystem health, water pollution is one of the worldwide issues that civilisation must solve in the twenty-first century. A range of water contaminants are produced by urbanisation, industrialisation, and climate change. In this chapter, we address some of the most recent and pertinent points on heavy metal releases, potential hazards to human health and the environment, and the materials and technology available for their removal. Here is several health risks associated with recurrent exposure to traces of heavy metals, such as arsenic, lead, cadmium, and mercury. We also discuss various methods for detecting heavy metals and the variables that may influence the removal of contaminants.

Keywords: *Pollution, metal ions, detection, removal, spectroscopic techniques*

INTRODUCTION

Environmental contamination constitutes a primary threat to our planet. Environmental pollution refers to the release of substances or energy into water, land, or air that results in or has the potential to result in acute (short-term) or chronic (long-term) harm to the earth's ecological equilibrium or diminishes the quality of life. Environmental pollution is the process of poisoning the environment, rendering it unfits for usage. Environmental pollution refers to the detrimental alterations in our environment that adversely impact flora, fauna, and humans. A material that induces pollution is referred to as a pollutant. Pollutants may exist as solid, liquid, or gaseous chemicals present at concentrations exceeding natural abundance, resulting from anthropogenic activity or natural events. Pollutants can inflict primary damage, resulting in direct and observable effects on the environment, or secondary damage, manifesting as subtle disruptions in the fragile equilibrium of the biological food web, which may only be discernible over extended timeframes. The industrialization of society, the advent of motorized vehicles, swift urbanization, population growth, exploitation of natural resources, and unregulated sewage and waste disposal from industries and urban areas have significantly escalated the generation of waste byproducts. Consequently, environmental pollution typically arises from energy transformations and resource utilization, resulting in the deposition of by-products in water, soil, or air. The issue of environmental pollution is significantly more extensive than commonly perceived. Environmental pollution is occurring globally, particularly in the forms of air and water contamination. China's cities, particularly the capital Beijing, exemplify air pollution, whereas India serves as a prime example of water pollution, notably with the contamination of the Ganges River. The most severe environmental contamination occurs in developing countries of the Third World due to their absence of sustainable management and basic sanitary facilities. Air and water pollution can lead to the mortality of several creatures within a given environment, including people. Environmental contamination has significant potential to impact human health (Fereidoun et al., 2007; Progressive Insurance, 2005). In the past thirty years, there has been a growing global apprehension regarding the public health effects associated with environmental pollution (Kimani, 2007). Human exposure to pollution is thought to be more severe currently than at any previous point in history (Schell et al, 2006). The NGO Pure Earth research indicates that one in seven deaths is attributable to pollution. A further analysis indicates that pollution causes 60% more fatalities than malaria, HIV/AIDS, and tuberculosis combined.

Classification of Pollutants: Pollutants are classified from several perspectives.

A. **Pollutants are classified into two categories based on their natural occurrence:**

    i. Quantitative Pollutants

    ii. Qualitative Pollutants

        i. **Quantitative Pollutants:** These compounds are naturally present in the environment but attain pollutant status when their concentrations rise owing to human activity. Carbon dioxide, when present in the atmosphere at concentrations exceeding normal levels due to anthropogenic activity, is categorized as a quantifiable pollutant.

        ii. **Qualitative Pollutants:** These are compounds that do not typically exist in nature but are introduced by humans, such as pesticides. Pollutants are classified into two categories based on their form after being discharged into the environment: (a) primary pollutants and (b) secondary pollutants.

- **Primary Pollutants:** These are substances released directly from their source and remain in the same form in which they were introduced into the environment. Such include ash, smoke, fumes, nitric oxide, sulfur dioxide, hydrocarbons, etc.

- **Secondary Pollutants:** These are pollutants that are generated from primary pollutants through chemical reactions with atmospheric constituents. Examples include sulfur trioxide, nitrogen dioxide, aldehydes, ketones, and ozone. Primary pollutants such as nitrogen oxides and hydrocarbons undergo reactions in sunlight, resulting in the formation of peroxyacyl nitrate (PAN) and ozone, which are classified as secondary pollutants.

B. **Pollutants can be categorized into two fundamental classes based on their natural disposal within ecosystems.**

i. **Biodegradable Pollutants:** These pollutants are rapidly decomposed through natural processes such as biological or microbial action, or through manmade systems like sewage treatment facilities. Degradable pollutants can be classified into two categories:

- **Rapidly degradable or non-persistent pollutant:** The degradation of these pollutants occurs at an accelerated rate. The degradation of sewage and organic waste from animals and plants occurs more rapidly.

- **Slowly degradable or persistent pollutant:** The degradation of these pollutants occurs at a significantly slower rate. The quantity of the pollutant appears to be constant over time. The breakdown of synthetic chemicals and radioactive elements such as Iodine-137 and Plutonium-239 requires an extended duration.

ii. **Non-degradable Pollutants:** These are compounds that eith\low rate in the natural environment. These encompass mercury salts, long-chain phenolic compounds, DDT, and aluminium cans, among others. Many of these mutants collect in the environment and undergo biological magnification as they progress through food chains in an undecomposed state. These may also interact with other environmental substances to generate poisons. These can be further categorized into two further classes:

- Waste: for instance, glass, plastic, phenolic materials, aluminium cans, etc.
- Toxins: for instance, radioactive materials, insecticides, atmospheric pollutants, and heavy metals.

Various pollutants, including heavy metals, microplastics, pesticides, and industrial contaminants, accumulate in the environment and cause long-term damage. Understanding these pollutants' sources, effects, and removal strategies is essential for developing sustainable solutions to mitigate pollution and protect natural resources. This chapter explores the impact of these major pollutants and discusses potential remediation techniques.

HEAVY METAL IONS

Heavy metal contamination originates from both natural and human activity. Mining, industrial effluents, electronic trash, and agricultural runoff substantially contribute to the environmental release of metals including lead (Pb), cadmium (Cd), arsenic (As), mercury (Hg), and chromium (Cr). These metals are non-biodegradable and remain in aquatic environments, soil, and atmosphere, resulting in bioaccumulation and biomagnification within the food chain. Metals such as lead, zinc, arsenic, copper, mercury, and cadmium in industrial wastewater negatively impact people and other animals. Arsenic contamination of groundwater has been documented in West Bengal, Orissa, Bihar, and Western Uttar Pradesh. Ingestion of arsenic-contaminated water results in the accumulation of arsenic in bodily tissues such as blood, nails, and hair, leading to skin lesions, roughness, dryness, thickening of the skin, and finally, skin cancer. Mercury pollution in aquatic environments induces Minamata illness in humans and dropsy in fish. Lead induces dyslexia, whereas cadmium toxicity results in Itai-Itai illness, among other conditions.

Table 1: Tolerance limits (ppm) in drinking water.

Metals	Limits	Metals	Limits	Metals	Limits
Al	1.0	Co	0.2	Li	5.0
Cd	0.001	Mn	2.0	Mo	0.005
Pb	5.0	Zn	5.0	Se	0.005
As	0.001	Ni	0.5	Hg	0.001
Cr	5.0	Cu	0.2	Be	0.50

Sources of heavy metal ions:

These heavy metals have been present in the Earth's crust since its genesis. The remarkable rise in the use of heavy metals has led to an impending growth of metallic substances in both terrestrial and aquatic environments. Heavy metal pollution has arisen from anthropogenic activities, primarily due to metal mining, smelting, foundries, and other metal-based industries, as well as the leaching of metals from various sources such as landfills, waste dumps, excretion, livestock and poultry manure, runoff, automobiles, and roadworks. The application of heavy metals in agriculture has been a secondary contributor to heavy metal pollution, including the use of herbicides, insecticides, and fertilizers. Natural phenomena can also elevate heavy metal contamination, including volcanic activity, metal corrosion, evaporation of metals from soil and water, sediment resuspension, soil erosion, and geological weathering.

Nearly 40% of the Earth's lakes and rivers have been contaminated by heavy metals (Zhou et al., 2020). The origins of these contaminants may be natural or manmade. Natural sources encompass interactions with metal-bearing rocks typically found in the environment and volcanic eruptions (Ali et al., 2019). Volcanic contributions may arise from substantial yet infrequent emissions, explosive volcanic events, or sustained low emissions, encompassing geothermal activity and degassing (Naggar et al., 2018). Anthropogenic sources encompass those linked to industrial activities (e.g., fossil fuel combustion, metal processing), agricultural practices (e.g., pesticides), and domestic actions (e.g., waste, detergents) (Gautam et al., 2014; Vareda et al., 2019).

Mining is a crucial sector to analyze, as it significantly influences the economies of both industrialized and developing nations. This activity produces a significant concentration of heavy metals, released from mineral extraction and conveyed via rivers and streams, where they may dissolve in water or exist within sediments (Vélez-Pérez et al., 2020). These metallic species often infiltrate groundwater, contribute to water scarcity, hinder crop growth due to soil erosion, and pose significant health risks to animals and local human populations (Birn et al., 2018). Human activities, urban development, and industrial growth are significant contributors to the rising levels of these pollutants, especially in recent decades. In India, the arsenic concentration in drinking water was estimated to be 107 µg/L (Roychowdhury et al., 2003). In Nigeria, elevated concentrations of Mn^{2+}, Pb^{2+}, Cr^{3+}, Ni^{2+}, Cd^{2+}, Ag^{+}, Hg^{2+}, As^{2+}, Se^{2+}, Zn^{2+}, and Co^{2+} were detected in water samples from various lead-zinc mining towns in Abakaliki (Obasi and Akudinobi, 2019). In Latin America, it was estimated that a minimum of 4.5 million individuals were chronically exposed to arsenic concentrations over 50 µg/L, with some instances reaching as high as 2000 µg/L (Fernández-Luqueño et al., 2013). In these three instances, the concentrations of the measured heavy metals significantly above the reference range established by the World Health Organization (WHO). Furthermore, about 140 million individuals across 50 nations consistently consume water with arsenic levels exceeding the WHO reference threshold of 10 µg/L (Bolisetty et al., 2019). High quantities of Cd^{2+}, Cu^{2+}, and Zn^{2+} were documented in nine coastal rivers in the Laizhou Bay basin in China. The highest concentrations detected exceeded the permissible limits established for potable water (Xu et al., 2017). Substantial mercury pollution was documented in the Coyuni River watershed, linked to the amalgamation procedures employed in artisanal gold mining in Venezuela (Sanchez et al., 2008). Multiple heavy metals were detected in the two principal rivers of Antalya, Turkey. The quantities of Sr and Al beyond the permissible limits established by the standard specification for reagent water ASTM D1193 (Leventeli et al., 2019). Considering the severity of the pollution issue, both emerging and developed nations must swiftly implement strategies to decrease the concentration of heavy metals in aquatic habitats and mitigate potential impacts on human health. In underdeveloped nations, a primary constraint is the limited economic capacity to develop and implement remediation solutions (Chowdhury et al., 2016).

Toxicological Effects

Heavy metals accumulate in their specific chemical forms or in conjunction with other metals, complicating their removal from water (Suhada et al., 2016). Trace amounts of these metals (e.g., Co, Cu, Cr, Fe, Mg, Mn, Mo, Ni, and Zn) are essential for metabolic functions and the prevention of illnesses and syndromes in humans (Tchounwou et al., 2012). Nonetheless, an elevated concentration of these heavy metals might result in significant human health issues, including muscular, musculoskeletal, and neurological degenerative disorders (Azeh Engwa et al., 2019). Prolonged exposure to hazardous heavy metals has been linked to muscular dystrophy, Alzheimer's disease, various cancers, and multiple sclerosis (Cabral Pinto et al., 2020; Cabral Pinto et al., 2020; Fu and Xi, 2020; Gerardo et al., 2020; Srivastava et al., 2017). Impairment or reduction of central neurological and cognitive functions, injury to the lungs, liver, kidneys, blood composition, and other vital organs has also been documented (Tan et al., 2018). The release of heavy metals into aquatic environments can result in physical, chemical, and biological disturbances, altering diversity, density, species composition, and community structure (Pratush et al., 2018; Ullah et al., 2017). The extent and form of alteration will mostly depend on the concentration of heavy metal ions and may disrupt enzymatic processes (Mokarram et al., 2020; Bobaker et al., 2019).

Detection

Prior to choosing an effective remediation method, it is essential to identify and quantify the concentration of heavy metals contaminating a specific water source. The chosen technique must be economical, environmentally sustainable, selective, and sufficiently sensitive to accurately identify trace amounts (Malik et al., 2019). Notwithstanding the challenges in fulfilling all these criteria, the advancement and integration of diverse techniques today permit the precise detection of heavy metals and the swift quantification of their amounts. This also facilitates the acquisition of knowledge regarding the extent of global water contamination issues. The subsequent sections delineate the spectroscopic, electrochemical, and optical methodologies employed for this objective.

Spectroscopic Methods

Heavy metals can be identified through highly sensitive and selective spectroscopic techniques, including atomic absorption spectroscopy (AAS), X-ray fluorescence (XRF), inductively coupled plasma mass spectrometry (ICP-MS), inductively coupled plasma optical emission spectroscopy (ICP-OES), neutron activation analysis, inductively coupled plasma atomic emission spectroscopy (ICP-AES), ion chromatography ultraviolet–visible spectroscopy (IC-UV Vis), total reflection X-ray fluorescence (TXRF), laser-induced breakdown spectroscopy (LIBS), atomic fluorescence spectroscopy (AFS), capillary electrophoresis (CE), and microprobes (MP). These approaches provide exceptionally low detection limits (femtomolar range) and facilitate the concurrent quantification of many heavy metals, which is a significant advantage (Zhong et al., 2016). AAS exhibits comparatively low detection limits of 2, 3, and 0.2 µg/L for Cu^{2+}, Pb^{2+}, and Cd^{2+} ions in water, respectively (Shirkhanloo and Mousavi, 2011). ICP-MS can simultaneously identify several metals in liquids, solids, and gases, offering exceptionally low detection limits ranging from ppb to ppt. When utilized with methods like laser ablation or chromatography, detection limits can be enhanced (Voica et al., 2012). Nonetheless, spectroscopic approaches necessitate intricate and costly apparatus, skilled staff often operating in a laboratory, and sample preparation methods that are generally challenging to stabilize (Kaur et al., 2016). The portable field versions of these techniques typically exhibit low sensitivity and accuracy (Kurup et al., 2017). For example, portable XRF equipment possesses significantly elevated detection limits, rendering it ineffective for trace detection (Kodom et al., 2012). Analysis of polluted water samples from Langshan, China, revealed that the minimum detectable concentrations using XRF were 21 ppm for Cu^{2+} and 28 ppm for Pb^{2+} (Zhou et al., 2018). XRF offers a significant advantage: the non-destructive investigation of both liquid and solid materials. Consequently, spectroscopic detection methods are employed in advanced laboratories possessing the requisite technical skills, which hinders this approach from being more cost-effective and environmentally sustainable.

Electrochemical Methods

Electrochemical techniques encompass potentiometry, amperometry, voltammetry, coulometry, impedance spectroscopy, and electrochemiluminescence (Pujol et al., 2014). The voltammetric and potentiometric methodologies have advanced significantly, leading to the development of linear sweep anodic stripping voltammetry (LSASV), square wave anodic stripping voltammetry (SWASV), differential pulse anodic stripping voltammetry (DPASV), cyclic cathodic voltammetry (CSV), cyclic voltammetry (CV), and chronopotentiometry (Chen et al., 2018; Kurup et al., 2017). Another area of research is biosensors, which, based on enzymes and other biological molecules or paired with nanostructured metallic materials, have enabled the detection of a greater number of pollutants in water samples (Gumpu et al., 2015). The methods employed with these techniques are comparatively straightforward and can be utilized with portable devices for in situ monitoring (Bansod et al., 2017). Electrochemical methods are cost-effective, user-friendly, dependable, and exhibit rapid analytical detection times. Nonetheless, these techniques have elevated detection limitations in comparison to spectroscopic and optical methods. Furthermore, they are not an effective alternative when other metal cations are present in the sample, as this diminishes detection sensitivity (Chen et al., 2018). Consequently, electrochemical techniques continue to face considerable problems regarding selectivity and sensitivity to achieve optimal detection efficacy.

Optical Methods

A range of optical sensors and test strips for the detection of heavy metals in aqueous solutions has been created. Techniques include the utilization of fluorophores or indicator dyes, biosensor assays, or the integration of an ionophore with a pH indicator (Ingale and Seela, 2012).

A further advancement involves utilizing nanoparticles and receptor molecules to ascertain the concentration of a singular analyte by the measurement of the sample's light absorbance (Berlina et al., 2019; Kurup et al., 2017). Investigations into portable optical enzymatic biosensors are significant, especially for the swift identification of contaminants in water samples (Lukyanenko et al., 2019). Numerous optical molecular detectors can ensure swift analysis; nevertheless, many of these methods are constrained by inadequate selectivity (Malik et al., 2019; Uglov et al., 2014). Although the optical method possesses certain advantages, its lack of selectivity may hinder decision-making and necessitates rectification to enhance this approach.

Removal of heavy Metals:

Wastewater treatment technologies are categorized into primary, secondary, and tertiary treatments. The primary treatment emphasizes the extraction of organic content and suspended solids from wastewater by physical and chemical methods.

Microfiltration, chemical filtration, coagulation, and flocculation are essential basic technologies for managing elevated quantities of heavy metals (Yenkie et al., 2019). Secondary treatments, whether anaerobic or aerobic, utilize natural microbes that transform organic and inorganic pollutants into simpler, safer chemicals, hence enhancing removal efficiency. The microbial efficacy in metal elimination is under investigation, however findings indicate significant removal rates (Bolisetty et al., 2019). Tertiary treatments encompass chemical oxidation, electrochemical precipitation, crystallization, distillation, photocatalysis, adsorption, membrane technologies, and ion exchange technologies. At this stage, wastewater is transformed into high-quality water. Removal efficiencies of up to 99% can be achieved when tertiary therapies are effectively integrated with primary and secondary treatments (Ince and Kaplan Ince, 2020). Each metal possesses distinct properties and behaviors; therefore, it is essential to examine the factors that affect their removal from water. The subsequent sections delineate the impact of pH, ionic strength, temperature, and natural organic matter concentration on the removal of heavy metals from domestic wastewater effluents, groundwater, rivers, and lakes.

REFERENCES

1. Abdel Maksoud, M.I.A., Elgarahy, A.M., Farrell, C., Al-Muhtaseb, A.H., Rooney, D.W., Osman, A.I., 2020. Insight on water remediation application using magnetic nanomaterials and biosorbents. *Coord. Chem. Rev.* 403, 213096.
2. Abdul Shukor, S.A., Hamzah, R., Abu Bakar, M., Noriman, N.Z., Al-Rashdi, A.A., Razlan, Z.M., Shahriman, A.B., Zunaidi, I., Khairunizam, W., 2019. Metal oxide and activated carbon as photocatalyst for waste water treatment. *IOP Conf. Ser. Mater. Sci. Eng.* 557.
3. Abdullah, N., Yusof, N., Lau, W.J., Jaafar, J., Ismail, A.F., 2019. Recent trends of heavy metal removal from water/wastewater by membrane technologies. *J. Ind. Eng. Chem.* 76, 17–38.
4. Abiaziem, C.V., Williams, A.B., Ibijoke, A., Vivian Abiaziem, C., Bassey Williams, A., Ibijoke Inegbenebor, A., Theresa Onwordi, C., Osereme Ehi Eromosele, C., Felicia Petrik, L., 2019. Adsorption of lead ion from aqueous solution unto cellulose nanocrystal from cassava peel. *J. Phys. Conf. Ser.* 1299, 012122.
5. Abu Qdais, H., Moussa, H., 2004. Removal of heavy metals from wastewater: A comparative study. *Desalination* 164, 105–110.
6. Ahmed, M.J.K., Ahmaruzzaman, M., 2016. A review on potential usage of industrial waste materials for binding heavy metal ions from aqueous solutions. *J. Water Process. Eng.* 10, 39–47.
7. Aksu, Z., Işoğlu, İ.A., Işoˇglu, I.A., 2005. Removal of copper(II) ions from aqueous solution by biosorption onto agricultural waste sugar beet pulp. *Process. Biochem.* 40, 3031–3044.
8. Al-Enezi, G., Hamoda, M.F., Fawzi, N., 2004. Ion exchange extraction of heavy metals from wastewater sludges. *J. Environ. Sci. Health-Part A Toxic/Hazardous Subst. Environ. Eng.* 39, 455–464.

9. Al-Saad, K.A., Amr, M.A., Atomic, E., Authority, E., Al-sulaiti, M., 2012. Iron oxide nanoparticles: applicability for heavy metal removal from contaminated water. *Arab J. Nucl. Sci. Appl.* 45, 335–347.

10. Alaba, P.A., Oladoja, N.A., Sani, Y.M., Ayodele, O.B., Mohammed, I.Y., Olupinla, S.F., Daud, W.M.W., 2018. Insight into wastewater decontamination using polymeric adsorbents. *J. Environ. Chem. Eng.* 6, 1651–1672.

11. Alharbi, N.S., Hu, B., Hayat, T., Rabah, S.O., Alsaedi, A., Zhuang, L., Wang, X., 2020. Efficient elimination of environmental pollutants through sorption reduction and photocatalytic degradation using nanomaterials. *Front. Chem. Sci. Eng.* 14, 1124–1135.

12. Ali, H., Khan, E., Ilahi, I., 2019. Environmental chemistry and ecotoxicology of hazardous heavy metals: Environmental persistence, toxicity, and bioaccumulation. *J. Chem.*

13. Almomani, F., Bhosale, R., Khraisheh, M., kumar, A., Almomani, T., 2020. Heavy metal ions removal from industrial wastewater using magnetic nanoparticles (MNP). *Appl. Surf. Sci.* 506, 144924.

14. Alothman, Z.A., Bahkali, A.H., Khiyami, M.A., Alfadul, S.M., Wabaidur, S.M., Alam, M., Alfarhan, B.Z., 2020. Low cost biosorbents from fungi for heavy metals removal from wastewater. *Sep. Sci. Technol.* 55, 1766–1775.

15. Amornwutiroj, S., Manpetch, P., Singhapong, W., Jaroenworaluck, A., 2019. Controllable synthesis of mesoporous magnetite / activated carbon composites as efficient adsorbents for hexavalent chromium removal. *J. Dispers. Sci. Technol.* 1–18.

16. Ao, W., Fu, J., Mao, X., Kang, Q., Ran, C., Liu, Y., Zhang, H., Gao, Z., Li, J., Liu, G., Dai, J., 2018. Microwave assisted preparation of activated carbon from biomass: A review. Renew. *Sustain. Energy Rev.* 92, 958–979.

17. Azimi, A., Azari, A., Rezakazemi, M., Ansarpour, M., 2017. Removal of heavy metals from industrial wastewaters: A review. *ChemBioEng. Rev.* 4, 37–59.

18. Azubuike, C.C., Chikere, C.B., Okpokwasili, G.C., 2016. Bioremediation techniques–classification based on site of application: principles, advantages, limitations and prospects. *World J. Microbiol. Biotechnol.* 32, 1–18.

19. Bagchi, S., Behera, M., 2020. Assessment of heavy metal removal in different bioelectrochemical systems: A review. *J. Hazard. Toxic. Radioact. Waste* 24.

20. Bansod, B., Kumar, T., Thakur, R., Rana, S., Singh, I., 2017. A review on various electrochemical techniques for heavy metal ions detection with different sensing platforms. *Biosens. Bioelectron.* 94, 443–455.

21. Barakat, M.A., 2011. New trends in removing heavy metals from industrial wastewater. *Arabian J. Chem.* 4, 361–377.

22. Barik, B., Kumar, A., Nayak, P.S., Achary, L.S.K., Rout, L., Dash, P., 2020. Ionic liquid assisted mesoporous silica-graphene oxide nanocomposite synthesis and its application for removal of heavy metal ions from water. *Mater. Chem. Phys.* 239, 122028.

23. Bashir, A., Ahmad, L., Sozia, M., Taniya, A., Mudasir, M., Bhat, A., 2019. Removal of heavy metal ions from aqueous system by ion-exchange and biosorption methods. *Environ. Chem. Lett.* 17, 729–754.

24. Berlina, A.N., Zherdev, A.V., Dzantiev, B.B., 2019. Progress in rapid optical assays for heavy metal ions based on the use of nanoparticles and receptor molecules. *Microchimica Acta.*

Chapter - 10

MONOSODIUM GLUTAMATE (MSG) AND ITS HEALTH IMPLICATIONS: A REVIEW

Ms. Prachi Singh, Ms. Neha Rajpoot, Ms. Ragni Singh and Dr. Fahad Ali*

Department of Zoology, D. S. College, Affiliated to Raja Mahendra Pratap, Aligarh-202001, India

Corresponding email: *sheikh.fa01@gmail.com

ABSTRACT

Monosodium glutamate (MSG) is a widely used food additive that enhances the flavor of food products. Generally, MSG is considered safe for consumption in moderate amounts (less than 1 g/day); however, some studies have reported adverse effects in certain animal species. This review aims to summarize the effects of MSG. Although several studies have detailed information on MSG, the primary objective of this review is to address gaps in the existing literature by summarizing both its beneficial and harmful effects as reported in previous research. A comprehensive search of recently published studies on MSG was conducted, including papers on its physical and chemical properties, structure, history, uses and harmful effects. The findings suggest that MSG consumption should be approached with caution. While it enhances the umami taste in foods, it has been shown to affect multiple organs, including the kidneys, liver, brain, testes, and ovaries. In conclusion, this review recommends natural alternatives such as turmeric, pot marigold, green tea, tomato, mango, and algae. These natural substances contain naturally occurring glutamate amino acids, provide flavor enhancement, and offer additional health benefits through antioxidant and anti-inflammatory properties, as well as modulation of tumor necrosis factors.

Keywords: *Monosodium glutamate, kidney, liver, antioxidant, anti-inflammatory, green tea, marigold, algae*

1. INTRODUCTION

Food provides nutrition and also supports social interaction. To improve food quality and shelf life, food additives are added to preserve freshness and enhance flavor, texture, and appearance. However, the increasing use of food additives has raised health concerns, as people are regularly exposed to these chemicals in their diets (Ramachandrappa et al., 2021). Common additives include flavor enhancers, preservatives, colorants, and stabilizers such as monosodium glutamate, ascorbic acid, sodium nitrate, benzoates, guar gum, gelatin, vinegar, and salt.

Monosodium glutamate (MSG) is one of the most widely used food additives in processed foods and imparts the characteristic umami taste (Kurihara, 2015). MSG is the sodium salt of glutamic acid, a non-essential amino acid, and is also known as China salt, sodium glutamate, or Ajinomoto. The European Union classifies MSG as a food additive permitted in limited quantities (Thuy et al., 2020). Glutamic acid was first discovered in 1866 by German chemist Karl Heinrich Ritthausen. In 1908, Kikunae Ikeda isolated it from kombu seaweed and identified it as the source of umami flavor. He later patented MSG production, and commercial manufacturing began in 1909 under the name Ajinomoto (Bera et al., 2017). Today, MSG consumption is highest in China, followed by Vietnam, Indonesia, and Japan (Thuy et al., 2020).

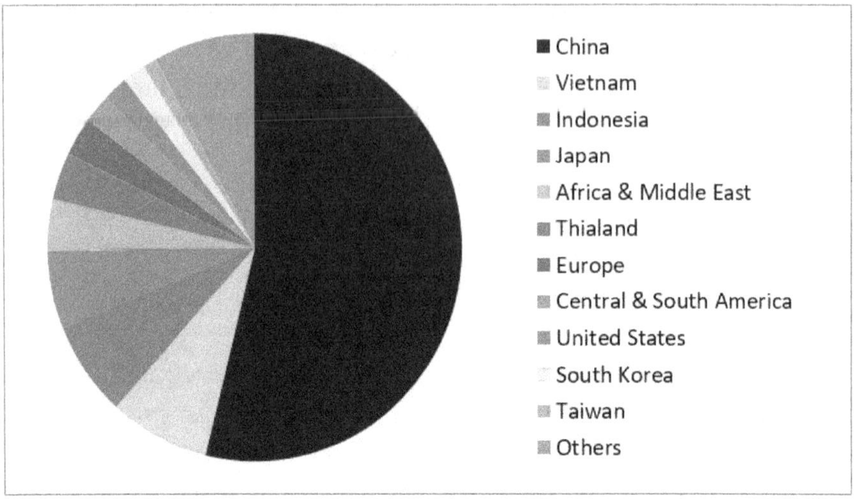

Figure 1: The figure shows the overall use of MSG during 2015- 2019 (Modified from Thuy et al., 2020).

The compound is usually available as the monohydrate, a white, odorless, crystalline powder (**Figure 2a**). The solid contains separate sodium cations Na+ and glutamate anions in zwitterionic form –OOCCH(NH+3) - (CH2)2-COO–. The chemical formula of MSG is $C_5H_8NO_4Na$ and its IUPAC name is sodium 2 amino-pentanedioate (Mahendra Dwivedi, 2024). Chemical structure of MSG is represented in Figure 2b. When dissolved in water, monosodium glutamate separates into Glutamate and Na ions. It readily dissolves in water but does not absorb moisture from the air and cannot be dissolved in Common organic solvents like ether. MSG remains stable during food processing and does not degrade while cooking. Similar to other amino acids, it undergoes a Maillard reaction (browning) when exposed to sugars at extremely high temperature.

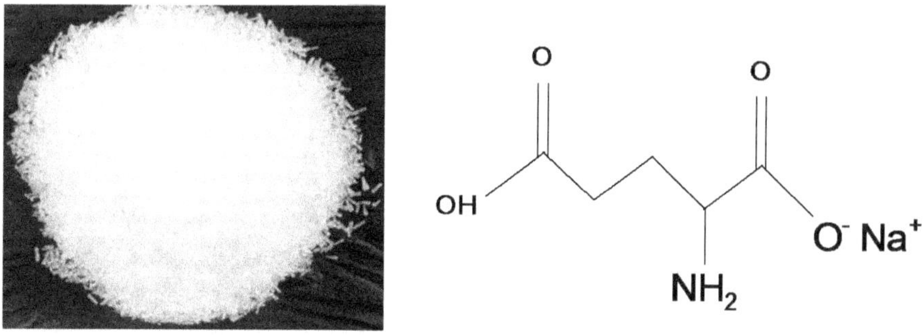

(a) (b)

Figure 2: (a) Crystals of Monosodium Glutamate (MSG); (b) Molecular structure of MSG.

2. SOURCES AND PRODUCTION OF MONOSODIUM GLUTAMATE

MSG is found naturally in substances such as hydrolyzed vegetable protein, autolyzed yeast, hydrolyzed yeast, yeast extract, soy extracts, and protein isolates. It also occurs naturally in tomatoes and cheeses (Marita *et al.*, 2022).

2.1. Manufacturing of Monosodium Glutamate

Monosodium glutamate (MSG) is now mainly produced through the fermentation of plant-based materials such as sugarcane, sugar beet, cassava, corn, or molasses, rather than being extracted from seaweed. This process is similar to the fermentation methods used in making yogurt, vinegar, and wine. In MSG production, glucose extracted from sugarcane is placed in a fermentation tank where microorganisms convert it into amino acids. The resulting solution is neutralized to form MSG, then decolorized, filtered, crystallized through evaporation, and dried to obtain the final MSG crystals (Mustafa et al., 2017).

3. APPLICATIONS OF MONOSODIUM GLUTAMATE IN VARIOUS FOOD PRODUCTS

I. **Fast Food:** MSG is commonly added to fast foods such as fried rice, chicken dishes, chowmein, and noodles to enhance flavor. Some large restaurant chains also use MSG to improve taste (Yang et al., 2023).

II. **Seasoning Blends:** MSG is used in seasoning mixes for foods like stews, tacos, and stir-fries to increase Savory and umami taste. It is also included in low-sodium products, bouillon cubes, and meat or fish seasonings to enhance flavour (Hwang et al., 2023).

III. **Milk Products and Frozen Meals:** MSG is found in certain dairy products like mozzarella and cream cheese, as well as frozen foods such as pizzas, macaroni and cheese, and ready meals to improve taste (Maluly et al., 2017).

IV. **Chips, Soups and Snack Foods:** Many snack foods including chips, corn snacks, and soups contain MSG to intensify their savory flavour (Thuy et al., 2020).

V. **Soft Tofu Stew Sauce and Vegetable Rice Porridge:** MSG is added to some soups and sauces, including tofu stew sauce and vegetable rice porridge, to enhance the Savory taste (Yehji Chung et al., 2022).

VI. **Processed Meats:** MSG may be present in processed meats such as sausages, hot dogs, deli meats, and jerky to enhance flavor and sometimes reduce sodium levels (Kos et al., 2023).

VII. **Condiments:** MSG is used in condiments like salad dressings, mayonnaise, ketchup, barbecue sauce and soy sauce to improve taste (Andarwulan et al., 2011).

4. HEALTH IMPLICATIONS ASSOCIATED WITH MSG INTAKE

A. **Oxidative Stress and Hepatotoxicity:** High intake of MSG can induce oxidative stress by increasing reactive oxygen species (ROS), leading to damage of cellular components such as lipids, proteins, carbohydrates, and nucleic acids. This disruption of cellular metabolism is associated with several disorders including neurodegenerative diseases, diabetes, and carcinogenesis (Abu-Taweel et al., 2014).

B. **Neurotoxicity:** MSG may overstimulate excitatory amino acid receptors, increase intracellular calcium levels and trigger enzymatic reactions that lead to neuronal cell death. Studies in animals have shown hypothalamic neuron damage and metabolic abnormalities such as obesity, insulin resistance, anxiety-like behavior, and inflammation (Niaz et al., 2018).

C. **Genotoxicity:** Several studies indicate that MSG has genotoxic and cytotoxic effects. Experiments using *Allium cepa* root tip cells and human peripheral blood lymphocytes showed chromosomal abnormalities and altered mitotic activity, indicating potential mutagenic effects (Swamy et al., 2013).

D. **Cardiovascular Diseases:** MSG is commonly used in instant noodles and processed foods with high sodium and low nutritional value. Frequent consumption has been associated with increased risk factors for cardiovascular diseases, including elevated blood glucose, cholesterol, triglycerides, and blood pressure (Hasenböhler et al., 2024).

E. **Reproductive Toxicity:** Animal studies show that MSG can induce oxidative stress in reproductive tissues, leading to testicular damage, reduced testosterone levels, decreased sperm count, and abnormalities in reproductive structures, potentially affecting male fertility (Oluwole et al., 2024).

Table 1. Table shows the effects of MSG administration in different animal models and in vitro cell cultures.

S.N	Disease	Dose/Day	Species	Effects	Reference
1.	Neurological disorder	410 fertilized chicken eggs divided into 5 group as control, distilled water, 0.12, 0.6 and 1.2 g MSG.	Babcock brown laying hens (*Gallus gallus domesticus*)	MSG could adversely affect embryonic growth and development along with causing necrosis of the neuron in developing brain, curling of legs and fingers.	Bolukbas *et al.*, 2023
2.	Hepato-toxicity and kidney damage	Dose of 0.6and 1.6 mg MSG / body weight for 14 days.	Adult rat	MSG is capable of producing alteration in body weight, liver and kidney because they are responsible for detoxification of foreign compound in body.	Manal Said Tawfik *et al.*, 2012
3.	Cardiovascular Disease	2mg/day MSG was administered sub-cutaneously during the first seven days	Newborn mice	MSG treated rats showed insulin resistance, higher arterial pressure. Latter is due to increased sympathetic neural output.	Konrad *et al.*, 2012

4.	Blood vascular disorder	3.33 & 6.66 mg MSG was given orally for 14 days.	Adult rabbit	It shows that MSG can increase the blood glucose level and cause diabetes mellitus in animals.	Oriaghan *et al.*, 2012
5.	Obesity	18 rats were divided into 3 groups having 6 in each. Group A (control), Group B (dose of 4mg/ body weight), Group C (2mg/ body weight).	Albino rat	In conclusion they find that MSG can cause weight gain at high dose.	Bunyan et al., 1976
6.	Lactation associated problem	Varying dose of 925mg, 180mg and 3700mg MSG respectively.	Female Wistar rats	Prolonged consumption of MSG could interfere with lactation associated function, mammary prolactin and oxytocin receptors.	Emmanuel *et al.*, 2024
7.	Chinese Restaurant Syndrome	Consumption of Chinese triple fried rice having MSG as main ingredient.	23 year old human male subject	Severe difficulty in swallowing saliva and inability to speak.	Bawaskar *et al.*, 2017

8.	Tissue damage	Dose of 5%, 1%, 0.2% & 0.04%	*Drosophila* (Fruit fly)	Significant pathological changes were found in male Drosophila administrated with 5% dose of MSG.	Kasozi *et al.*, 2018
9.	Hepato-cardiac disease	600 mg/kg	Male rats	Result reveals rise in body weight, dyslipidemia, inflammation, and hepato-cardiac toxicity.	Banerjee *et al.*, 2021
10.	Muscle pain and headache	Dose of 150mg MSG diluted in 400ml of soda.	12 Temporo-mandibular (TMD) joint disorder patients and 12 healthy human subjects participated.	They suggest that TMD patients may be particularly sensitive to MSG compare to healthy ones.	Shimada *et al.*, 2015
11.	Growth retardation	10 test concentrations of MSG of 10, 30, 50, 100, 150, 200, 250, 300, 400, 500 mg/l were used as dosage.	Zebra fish (*Danio rerio*)	There were distinguishable negative alterations such as growth retardation, shrinkage of chorion, lack of pigmentation, tail deformities and scoliosis in embryos.	Jadav, 2019

12.	Damage to cell organelles	6 g dose of MSG	Adult male rat *Rattus norvagicus*	Severe distortion of hepatocytes, swollen mitochondria and vesiculated ER with picknotic nuclei.	El-Meghawry *et al.*, 2012
13.	Seizures and Brain Damage	Use of AuxiGro in dog food contains hydrolyzed proteins and MSG.	Dog	Found that MSG in dog foods leads to seizures in dogs.	Scher *et al.*, 1992
14.	Phagocytic activity of MSG on a blood neutrophils	15 mice injected with MSG (G Group) other 8 serve as control (C Group)	Newborn mice	The neonatal MSG treatment revealed decrease in blood lymphocytes and obesity	Hriscu *et al.*, 1997
15	Anorexia	4 broths- chicken broth, broth with 1.19 MSG and nucleotides, broth with 1.22g MSG broth with added fat were used as dose.	35 normal weight women aged 20-40 years volunteered in the experiment	Supplementing chicken broth with MSG reduces hunger and desire to snack but does not affect energy intake in women.	Carter *et al.*, 2011

17.	Nervous tissue damage	Broiler chicks divided into 4 groups and 0.01, 0.25, 0.5 and 1.0% MSG respectively mixed in food.	Lohman Broiler chicks	Histopathological evidences showed lesions in mid brain and cerebellum. Neuronal gliosis, chromatolysis, and angulation in the brain were also observed.	Khadiga *et al.*, 2009
18.	Growth retardation and vomiting	Dose of 0, 4.8, 9.1 and 16.7% MSG was given to infant squirrel monkeys.	Infant squirrel monkeys	Monkeys receiving largest dose of MSG showed retardation and elevated circulating glutamate.	Wen *et al.*, 1973
19.	Nuclear damage	92 adult male and female Siberian hamsters are treated with variable doses of MSG.	Siberian hamsters (*Phodopus sungorus)*	MSG induced arcuate nucleus damage affects both natural torpor and 2DG-Induced torpor like hypothermia in Siberian hamsters.	Pelz *et al.*, 2008
20.	Cochleo-toxic	30 Guinea pig divided into control and intervention groups (100 mg MSG/ day; 300 mg MSG/ day).	Guinea pig	In MSG groups moderate to severe degeneration and cell loss in outer hair, support cells and partial loss of outer hair cell stereocilia was noted.	Guven *et al.*, 2022

21.	Oxidative stress and physiological imbalances	Total 300 laying hens were divided into 6 groups. Each group was given an oral dose of 0.00, 0.25, 0.50, 0.75, 1.00 and 1.25g of MSG *ad libitum*	Laying Hens	Results showed that at high dose of MSG adversely affects serum electrolytes and antioxidant status of laying hens.	Olarotimi, 2020
22	Nephro-toxicity and pro-oncogenic effect of MSG	134 quails received a daily dose of 10g, 30g, 50g MSG/kg respectively for 30 days.	Quail (*Coturnis japonica)*	In kidney of birds from 30-50g there is degradation of cortical and medullar cyto-architecture and also necrosis of PCT and DCT.	Solcan *et al.*, 2018

5. MECHANISM OF TASTE PRODUCTION BY MSG

When ingested, MSG dissociates into sodium and glutamate ions, the glutamate ion bind and work as ligand for activating specific T1R1/T1R3 glutamate receptors on the taste buds of tongue which trigger the taste sensation and then send signals to the brain, resulting in the perception of the "umami" taste, essentially enhancing the savory flavor of food by interacting with these dedicated taste receptors on the tongue (Yamamoto *et al.*, 2023). This mechanism is considered safe by most regulatory agencies as the glutamate is a naturally occurring amino acid found in many foods and this function is performed by Monosodium glutamate when it is consumed in a limiting amount.

6. THERAPEUTIC POTENTIAL OF VARIOUS PROTECTANTS USED AGAINST MSG

Several natural substances have shown protective effects against **MSG-induced toxicity** due to their antioxidant and anti-inflammatory properties.

1. ***Calendula officinalis* (Pot marigold):** Used in teas and herbal extracts; possesses medicinal and protective properties (Shivasharan et al., 2013).
2. ***Cucurbita ficifolia* (Fig leaf gourd):** Reduces body weight and inflammation in MSG-induced obesity models by regulating inflammatory markers such as TNF-α and IL-6 (Fortis-Barrera et al., 2017).
3. ***Curcuma longa* (Turmeric):** Curcumin protects against MSG-induced neurotoxicity, oxidative stress, and reproductive damage due to its strong anti-inflammatory and antioxidant effects (Hajihasani et al., 2020).
4. ***Hibiscus sabdariffa*:** Used in beverages and herbal medicine; has potential health-protective properties (Alharbi, 2024; Hussaini, 2024; Alsnami, 2024).

5. ***Camellia sinensis* (Green tea):** Rich in polyphenols and antioxidants that reduce oxidative stress and liver toxicity caused by MSG (Kasmara et al., 2025).
6. ***Mangifera indica* (Mango):** Mango extracts and peels contain antioxidants that help reduce MSG-induced toxicity (Egbuonu, 2022).
7. ***Solanum lycopersicum* (Tomato):** Tomato seeds and extracts possess antioxidant properties useful for protecting against oxidative damage (Skypala et al., 2021).
8. ***Zingiber officinale* (Ginger):** Exhibits antioxidant and anti-inflammatory activities that help reduce MSG-induced oxidative stress (Hussain et al., 2017).
9. **Walnut:** Contains natural glutamate and beneficial nutrients that may counter MSG effects (Uti et al., 2020).
10. ***Ginkgo biloba*:** Shows neuroprotective and reproductive protective effects against MSG toxicity (AboHashish et al., 2015).
11. **Vitamins (C, D, E, L-arginine):** Help protect liver and kidney tissues and reduce oxidative damage caused by MSG (Ayman et al., 2013).
12. **Algae (Spirulina, Chlorella):** Rich in antioxidants that neutralize free radicals and protect against MSG-induced organ damage (Trotta et al., 2024).

CONCLUSION

The utilization of monosodium glutamate (MSG) is presently a subject of considerable debate. Although the U.S. Food and Drug Administration (FDA), alongside other health organizations such as the World Health Organization (WHO) and the European Food Safety Authority (EFSA), classify MSG as "generally safe for consumption," some researchers have identified its detrimental effects on various bodily organs at different dosages. Consequently, our review concludes that, despite MSG's function as a flavor enhancer that improves the palatability of food, it is prudent to avoid its consumption. MSG has been associated with a spectrum of diseases and disorders, including Chinese Restaurant Syndrome (CRS), genotoxicity, reproductive toxicity, neurological disorders, hepatotoxicity, and oxidative stress. To address this concern, the optimal solution is to employ natural products such as turmeric, pot marigold, green tea, tomato, and algae. These natural alternatives not only serve as flavor enhancers but also provide natural sources of glutamic acid. Therefore, future recommendations in this domain include the effective utilization and extensive research on the benefits of natural protectants to enhance their use as substitutes for MSG.

REFERENCES

1. AboHashish, N. A. & Elbastawesy, S. M. (2023). The nephrotoxic effect of chloroquine, the off-label anti covid 19 and possible protective role of *Ginkgo biloba* extract in male albino rats. 23(1).
2. Abu-Taweel, G. M. et al., (2014). Cognitive and biochemical effects of monosodium glutamate and aspartame, administered individually and in combination in male albino mice. *Neurotoxicology and Teratology*, *42*, 60-67.
3. Alharbi, A. E. et al., (2024). A Comprehensive Review of the Antimicrobial Effects of Hibiscus Species. *Cureus, 16*(11).
4. Andarwulan, N. et al., (2011). Free glutamate content of condiment and seasonings and their intake in Bogor and Jakarta, Indonesia. *Food and Nutrition Sciences, 2*(7), 764-769.
5. Ayman, E. L. et al., (2013). The effect of vitamin C administration on monosodium glutamate induced liver injury. An experimental study. *Experimental and toxicologic pathology, 65*(5), 513-521.
6. Banerjee, A. et al., (2021). Monosodium glutamate causes hepato-cardiac derangement in male rats. *Human & experimental toxicology, 40*(12_suppl), S359- S369.
7. Bawaskar, H. S. et al., (2017). Chinese restaurant syndrome. *Indian Journal of Critical Care Medicine*: Peer-reviewed, Official Publication of Indian Society of Critical Care Medicine, *21*(1), 49.

8. Bera, T. K. et al., (2017). Effects of monosodium glutamate on human health: A systematic review. *World journal of pharmaceutical sciences, 5*(5), 139-144.

9. Bolukbas, F. & Oznurlu, Y. (2023). Determining the effects of in ovo administration of monosodium glutamate on the embryonic development of brain in chickens. *Neurotoxicology, 94*, 87–97.

10. Bunyan, J. et al., (1976). The induction of obesity in rodents by means of monosodium glutamate. *British Journal of Nutrition, 35*(1), 25-39.

11. Carter, B. E. et al., (2011). Supplementing chicken broth with monosodium glutamate reduces hunger and desire to snack but does not affect energy intake in women. *British Journal of Nutrition, 106*(9), 1441-1448.

12. Chung, Y. et al., (2022). Effect of Monosodium Glutamate on Salt and Sugar Content Reduction in Cooked Foods for the Sensory Characteristics and Consumer Acceptability. *Foods, 19*;11(16):2512.

13. Dwivedi, M.(2024). The Science behind monosodium glutamate: Flavor modulation, Food palatability and potential health effect. *Southeast Asian Journal of Case REPORT and Review, 11*(3) 69-69.

14. Egbuonu, A. C. C. (2022). Antimicrobial properties and effect of ethanol extract of Mango seed kernel on serum electrolytes of monosodium glutamate-challenged rats. *Journal of Chemical Society of Nigeria,* 37(2),128-132.

15. El-Meghawry, E. M. et al., (2012). The effect of vitamin C administration on monosodium glutamate induced liver injury: An experimental study. *Experimental and toxicologic pathology, 64*(7-8),513-521.

16. Emmanuel, N. S. et al., (2024). Preliminary monosodium glutamate-induced changes in mammary gland receptors and gene expression, Water channel, oxidative stress, and some lactogenic biomarkers in lactating rats. *The Journal of Basic and Applied Zoology, 85*(1), 3

17. Eweka, A. O. et al., (2010). Histological studies of the effects of monosodium glutamate of the fallopian tubes of adult female Wistar rats. *North American journal of medical sciences, 2*(3), 146.

18. Fortis-Barrera, Á. et al., (2017). *Cucurbita ficifolia* (Cucurbitaceae) modulates inflammatory cytokines and IFN-γ in obese mice. *Canadian journal of physiology and pharmacology, 95*(2), 170-177.

19. Gottardo, F. M. et al., (2022). Use of monosodium glutamate in foods: the good, the bad, and the controversial side. *ABCS Health Sciences, 47*, e022305-e022305.

20. Guven, S. G. et al., (2022). Does Oral Monosodium Glutamate Have a Cochleotoxic Effect? An Experimental Study. *Audiology and Neurotology, 27*(2), 109-121.

21. Hajihasani, M. M. et al., (2020). Natural products as safeguards against monosodium glutamate-induced toxicity. *Iranian journal of basic medical sciences, 23*(4), 416–430.

22. Hasenböhler, A. et al., (2024). Food additive monosodium glutamate and risk of cardiovascular diseases-NutriNet-santé cohort. *European Journal of Public Health, 34*(Supplement_3), 144-234.

23. Hriscu, M. et al., (1997). Effects of monosodium glutamate on blood neutrophils phagocytic activity and phagocytic response in mice. Romanian Journal of Physiology: *Physiological Sciences, 34*(1-4), 95-101.

24. Hussain, U. K. et al., (2017). Ginger and propolis exert neuroprotective effects against monosodium glutamate-induced neurotoxicity in rats. *Molecules, 22*(11), 1928.

25. Kos, I. et al., (2023). Glutamate in meat processing–origin, function and novel application. *Journal of Central European Agriculture, 24*(3), 624-633.

26. Jadhav, H. S. (2019). Investigating the effects of monosodium glutamate using zebrafish as an animal model (Doctoral dissertation, Indian Institute of Technology Hyderabad).

27. Ji-sun Hwang et al., (2025). The role of MSG and nucleotide-based seasonings in Doenjang soup, *Food Quality and Preference, 131*, 105579.

28. Kasmara, D. P. et al., (2025). Mini review of plant products as food supplement against MSG-induced liver injury: antioxidant, oxidative stress and histological prospects. *Frontiers in Pharmacology, 16*, 1522814.

29. Kasozi, K. I. et al., (2018). Low concentrations of monosodium glutamate (MSG) are safe in male Drosophila melanogaster. *BMC Research Notes, 11*, 1-5.

30. Khadiga, K. et al., (2009). Response of broiler chicks to dietary monosodium glutamate. *Pakistan Veterinary Journal, 29*(4), 165-168.

31. Konrad, S. P. et al., (2012). Monosodium glutamate neonatal treatment induces cardiovascular autonomic function changes in rodents. *Clinics, 67*, 1209-1214.

32. Kos, I. et al., (2023). Glutamate in meat processing–origin, function and novel application. *Journal of Central European Agriculture, 24*(3), 624-633.

33. Kurihara, K. (2015). Umami the fifth basic taste: history of studies on receptor mechanisms and role as a food flavor. *BioMed research international, 1*, 189402.

34. Maluly, H. D. et al., (2017). Monosodium glutamate as a tool to reduce sodium in foodstuffs: Technological and safety aspects. *Food science & nutrition, 5*(6), 1039-1048.

35. Mustafa, Z. et al., (2017). Monosodium glutamate, commercial production, positive and negative effects on human body and remedies-a review. *IJSRST,3*(4), 425-435.

36. Niaz, K. et al., (2018). Extensive use of monosodium glutamate: A threat to public health? *EXCLI jourbody19(17)*, 273-278.

37. Olarotimi, O. J. (2020). Serum electrolyte balance and antioxidant status of broiler chickens fed diets containing varied levels of

monosodium glutamate (MSG). *Bulletin of the National Research Centre, 44*, 1-7.

38. Oluwole, D. T. et al., (2024). Disruptive consequences of monosodium glutamate on male reproductive function: A review. *Current Research in Toxicology, 6,* 100148.

39. Oriaghan, E. A. et al., (2012). The effect of monosodium glutamate (MSG) on blood glucose in adult rabbits as models. International Journal of Basic, *Applied and Innovative Research, 1*(1), 10-18.

40. Pelz, K. M. et al., (2008). Monosodium glutamate-induced arcuate nucleus damage affects both natural torpor and 2DG-induced torpor-like hypothermia in Siberian hamsters. *American Journal of Physiology- Regulatory, Integrative and Comparative Physiology, 294*(1), R255-R265.

41. Ramachandrappa, N. K. et al., (2021). Safety of cooked foods in home and street vended conditions beyond two hours: relevance of safe storage temperature message from WHO's five keys for safe food in India. International Journal of Food Safety, *Nutrition and Public Health, 6*(2), 104-119.

42. Scher, W. & Scher, B. M. (1992). A possible role for nitric oxide in glutamate (MSG)-induced Chinese restaurant syndrome, glutamate-induced asthma, 'hot-dog headache', pugilistic Alzheimer's disease, and other disorders. *Medical hypotheses, 38*(3), 185-188.

43. Shimada, A. et al., (2015). Differential effects of repetitive oral administration of monosodium glutamate on interstitial glutamate concentration and muscle pain sensitivity. *Nutrition, 31*(2), 315-323.

44. Shivasharan, B. D. et al., (2013). Protective effect of Calendula officinalis L. flowers against monosodium glutamate induced oxidative stress and excitotoxic brain damage in rats. *Indian Journal of Clinical Biochemistry, 28*, 292-298.

45. Skypala, I. J. et al., (2021). Non-specific Lipid Transfer Protein Allergy Across Europe. Non-specific lipid-transfer proteins: Allergen structure and function, cross-reactivity, sensitization, and epidemiology. *Clinical and Translational Allergy, 11*(3), e12010.

46. Solcan, C. et al., (2018). Nephrotoxic and prooncogenic effects of monosodium glutamate in quail (*Coturnix japonica*). *Pakistan Veterinary Journal, 38*, 29-34.

47. Swamy, A. V. et al., (2013). Neuroprotective activity of pongamia pinnata in monosodium glutamate-induced neurotoxicity in rats. *Indian journal of pharmaceutical sciences, 75*(6), 657.

48. Tawfik, M. S. & Al-Badr, N. (2012). Adverse effects of monosodium glutamate on liver and kidney functions in adult rats and potential protective effect of vitamins C and E. *Food and Nutrition Sciences, 3*(5), 651.

49. Thuy, L. N. (2020). A mini review about monosodium glutamate. *Bulletin UASVM Food Science and Technology, 77*(1), 1-12.

50. Trotta, F. et al., (2024). Advancing food preservation: sustainable green-AgNPs bionanocomposites in paper-starch flexible packaging for prolonged shelf life. *Polymers, 16*(7), 941.

51. Uti, D. E. et al., (2020). African walnuts attenuate ectopic fat Uti, D. and associated peroxidation and oxidative stress in monosodium glutamate-obese Wistar rats. *Biomedicine & pharmacotherapy, 124*, 109879.

52. Wen, C. P. et al., (1973). Effects of dietary supplementation of monosodium glutamate on infant monkeys, weanling rats, and suckling mice. *The American journal of clinical nutrition, 26*(8), 803-813.

53. Yamamoto, T. & Inui-Yamamoto, C. (2023). The flavor-enhancing action of glutamate and its mechanism involving the notion of kokumi. *National Partner Journal Science of Food, 7*(1), 3.

54. Yang, L. et al., (2023). A multifaceted review of monosodium glutamate effects on human health and its natural remedies. *Food Materials Research, 3*(1), 16.

Chapter - 11

GEOLOGICAL APPLICATION IN ECOTOXICOLOGY: PATHWAYS TO ENVIRONMENTAL SUSTAINABILITY

Rajan Shukla[1*] and Dr. Ainul H. Ansari[2]

[1]Department of Geology, D. S. College, G. T. Road, Aligarh, India, 202001

[2]Department of Geology, Aligarh Muslim University, Aligarh, India, 202002

***Corresponding email: rajanshukladsc@gmail.com**

ABSTRACT

This chapter explores the integrative role of geological science in advancing ecotoxicology and supporting environmental sustainability. It focuses on how geological factors such as mineral composition, soil stratification and hydrological dynamics influence the mobility, transformation, and toxicity of environmental pollutants. A mixed-methods approach was employed, combining field-based geological sampling, laboratory ecotoxicological assays, and geostatistical analysis across diverse ecological settings. Findings indicate that lithological variations significantly influence contaminant bioavailability and toxicity. Delayed ecotoxic responses observed in sediment-bound pollutants highlight the necessity of incorporating geological matrices in ecological risk assessments. The study underscores the value of integrating geochemical indicators to improve predictions of pollutant behavior and ecosystem vulnerability.

Practically, this research offers a framework for enhancing environmental monitoring programs, especially in geologically complex regions. It delivers actionable insights for policymakers, environmental planners and remediation professionals engaged in designing localized sustainability strategies. The chapter's originality lies in its interdisciplinary approach, bridging geology and ecotoxicology to foster a predictive and adaptive model for environmental management. This shift from reactive assessment to proactive stewardship strengthens ecological resilience and promotes informed, sustainable land-use practices.

Keywords: *Geological ecotoxicology, environmental sustainability, pollutant bioavailability, geochemical analysis, sustainable remediation*

1. INTRODUCTION

Ecotoxicology, which has its roots in the fields of toxicology and ecology, is an advancing field that addresses the effects of toxic substances on the environment and its resident biota. Although conventional ecotoxicological analyses usually focus on chemical and biological processes, the geochemical basis of ecosystems has not been fully addressed. Soils, water and other geological materials are responsible for the transport, distribution, and bioavailability of contaminants. This may result in poor environmental risk assessments and, hence, poor remediation decisions (Zhu & Anderson, 2023).

The interface between geological sciences and ecotoxicology holds out a prospect for environmental sustainability. Substrates are geological features that not only control the movement of contaminants through absorption, desorption and ion exchange but also control the storage or transformation of pollutants in sediments and aquifers for the long term. These processes are important to know to identify contaminant hot spots, develop predictive models, and design site-specific remediation plans. This becomes particularly important in areas prone to mining, industrial pollution, or agrochemical use, where geological knowledge, when combined with ecotoxicological analysis, enhances the diagnostic capacity and environmental decision-making (Hudson-Edwards et al., 2023a; Appelo & Postma, 2023).

- Assessing how lithological and hydrogeological features affect contaminant mobility and toxicity.
- Identifying geochemical indicators that correlate with ecotoxic responses.
- Evaluating current modeling frameworks that integrate geological data into ecological risk assessment.

The scope encompasses terrestrial and aquatic systems, with a focus on both natural and anthropogenically impacted environments.

Key chapter questions addressed in this study include:

- How do geological variables affect the persistence and bioavailability of pollutants?
- What role do mineralogical and geochemical characteristics play in modulating ecotoxicological outcomes?
- Can integrating geological insights improve the accuracy of environmental risk predictions?

The core hypothesis is that geological conditions significantly influence ecotoxicological responses, and their inclusion in environmental assessments leads to more effective and region-specific sustainability strategies (Choppola et al., 2023).

2. LITERATURE REVIEW

Ecotoxicology is a science that has developed from a chemistry-based field to an ecology, microbiology, and hydrology-oriented field within the last 50 years. In step with the growing concern about the sustainability of the environment, it has become a broad interdisciplinary science that incorporates geosciences in the studies of heavy metals, mining activities, and soil degradation (Hudson-Edwards et al., 2023a; Leahy and Colwell, 2023). This is because the interactive process between the geological media and biological systems is more complex than can be understood from a geoscience perspective, and this interdisciplinary development is, therefore, vital.

Soil and water are the natural sources of metals and other contaminants, which, after entering the food chain, may pose a risk to the biota. The mineralogical constituents of the soil, the grain size, porosity, and the dynamics of the groundwater determine the movement, reaction, and fixation of pollutants like arsenic, mercury, lead, and other persistent organic substances (Canfield, 2023; Merkel et al., 2023). Some of these processes include sorption onto clay minerals or redox transformations that can either diminish or enhance the toxicity of the contaminants. However, these factors are sometimes simplified or omitted in risk assessments and thus result in the underestimation of risks to the environment.

The use of geochemical indices such as pH, electrical conductivity, redox potential, and heavy metal speciation is useful in assessing ecotoxicological consequences (Al-Battashi et al., 2023; Liu et al., 2023). These parameters are not only warning signs but also help in determining the sensitivity of ecosystems through geo-bio signatures. Changes in enzymatic activities in aquatic organisms or oxidative stress in plants are also associated with higher metal concentrations, especially in catchments with sediment-dominated inputs (Rajala et al., 2023; Gadd, 2023).

Reactive transport models and spatial risk prediction tools have been created to model pollutant behavior in geological matrices. However, many of these models are deficient in biological integration or site-specific geological calibration. Some of the tools, like PHREEQC and MODFLOW, are good in hydro-geochemistry but do not have ecotoxicological feedback. Therefore, their applicability is rather restricted for actual conditions of environmental management tasks.

Despite the expanding spectrum of knowledge in the geo-ecotoxicological borderline, there are still many questions unanswered. First, there are an insufficient number of studies that are region-specific and incorporate specific geological features such as basalt, karst, or glacial systems. Second, there is a deficiency of integrated databases that link geological and ecotoxicological data for large-scale comparison (Palmer et al., 2023a; Palmer et al., 2023b). Lastly, although the microbial processes associated with geochemical cycling are now receiving more attention, the link between the latter and pollutant toxicity in geological environments is still largely unexplored (Ma et al., 2024; Cheung and Gu, 2023).

Thematic Organization: Chapters in this domain are grouped along thematic clusters, which include heavy metal behavior in mining-affected areas, soil contamination and remediation, and groundwater pollution under different geogenic conditions.

Critical Analysis: Although many studies have significantly advanced the understanding of environmental contaminants, a considerable number are still conducted under controlled or semi-natural conditions, often relying on generalized assumptions regarding soil or aquifer properties. Such limitations may reduce the applicability of these findings to diverse real-world environments. Additionally, insufficient collaboration between geochemists and ecotoxicologists often results in compartmentalized approaches to environmental assessment, hindering comprehensive interpretations (Oberle et al., 2023; Rawlings, 2023).

The purpose of this study is to try to bridge the gap between disciplines by combining geological modeling with ecotoxicological indicators in a unified framework. In this way, the chapter gaps identified above can help to build a predictive, site-specific, and geologically informed approach to environmental sustainability.

3. METHODOLOGY

3.1 Study Area and Geological Context

The study was conducted across three ecologically diverse regions:

1. A mining-intensive zone with sulfide-rich bedrock,
2. An agricultural basin characterized by loamy soils and high groundwater interaction, and
3. A coastal plain with saline intrusion vulnerability.

These sites were chosen to determine the impacts of different geological substrates on pollutant transport and ecological effects. The lithologies of the underlying rocks were granitic, lateritic and alluvial, respectively (Zhu & Anderson, 2023).

Table 1. Geological Characterization of Study Sites

Site Code	Region Type	Dominant Lithology	Soil Type	Groundwater Table Depth (m)	Known Pollutants
S1	Mining Zone	Sulfide Bedrock	Clay-Silt Mix	12.4	Arsenic, Lead
S2	Agricultural Zone	Loamy Alluvium	Sandy Loam	4.7	Nitrates, Pesticides
S3	Coastal Zone	Sedimentary	Saline Clay	1.2	Mercury, Chlorides

3.2 Sampling Design and Environmental Matrix Selection

In order to guarantee the representation from different geological and contamination profiles, a stratified purposive sampling technique was applied. Soil sediment, groundwater, and surface water samples were collected in triplicate per location in both the dry and wet seasons. The sampling depths were 0–30 cm for soil and 1.5–15 m for groundwater (Oelkers & Schott, 2023).

Table 2. Environmental Matrices and Sampling Protocols

Matrix	Depth Range	Collection Tool	Preservative Used	Holding Time Max (hrs)
Soil	0–30 cm	Stainless Steel Auger	Dry ice	48
Sediment	0–10 cm	PVC Core Sampler	Acid-washed container	24

Groundwater	1.5–15 m	Submersible Pump	HNO2 (for metals)	12
Surface Water	Surface–0.5 m	Grab Sampler	Ice-packed container	6

3.3 Analytical Techniques and Instrumentation

All samples underwent both geochemical and toxicological characterization using standardized laboratory protocols.

- **Heavy metals** were analyzed using **Inductively Coupled Plasma Mass Spectrometry (ICP-MS)**.
- **Organic pollutants** were determined via **Gas Chromatography-Mass Spectrometry (GC-MS)**.
- **Toxicity bioassays** were performed on *Daphnia magna* and *Lemna minor* following OECD guidelines.

Table 3. Laboratory Techniques and Instrumentation

Parameter	Instrument/ Method	Detection Limit	Calibration Method
Heavy Metals (As, Pb, Hg)	ICP-MS	0.1 µg/L	External standards
Nitrate and Phosphate	UV-Vis Spectrophotometry	0.01 mg/L	Calibration curve (R^2>0.99)
VOCs and PAHs	GC-MS	0.05 µg/L	Split/splitless injection
pH, EC, TDS	Multiparameter Probe	-	Daily 3-point calibration

Safety protocols were maintained during sample processing, including fume hood use, PPE compliance, and weekly equipment calibration.

3.4 Risk Assessment Models and Toxicological Metrics

To evaluate ecological risk, a **tiered assessment model** was adopted combining:

- **Hazard Quotient (HQ)**
- **Ecological Risk Index (ERI)**
- **Bioaccumulation Factor (BAF)**

Benchmark values were sourced from WHO and USEPA standards. Additionally, sediment quality guidelines were used to assess toxicity thresholds.

Table 4. Risk Assessment Metrics and Thresholds

Metric	Formula	Risk Threshold
HQ	C_e/C_t	>1 = Risk
ERI	$\Sigma(T_i \times C_i/B_i)$	>40 = Moderate Risk
BAF	$C_{or}g^{anism}$ / C_w_{ater}	>1000 = Bioaccumulative

Where C_e is exposure concentration, C_t is toxicity reference, T_i is toxicity response factor, and B_i is background level.

3.5 Statistical and Geospatial Data Analysis Approaches

Various statistical analysis tools, including Principal Component Analysis (PCA) and Cluster Analysis, were applied to detect the contaminant sources and geochemical interactions. ArcGIS Pro was used for spatial analysis and pollution plume modeling (Oelkers & Schott, 2023).

Table 5. Statistical and Geospatial Tools

Analysis Type	Software Used	Purpose
PCA	RStudio (FactoMineR)	Identify variance contributors
Cluster Analysis	IBM SPSS	Group similar pollution patterns
Geospatial Mapping	ArcGIS Pro	Visualize pollutant distribution
Heatmaps	QGIS	Hotspot identification

System Design/Configuration and Setup Steps

The study followed a **modular experimental design** where each region was configured into a geochemical risk framework:

1. **Sample pre-processing** under standardized temperature and humidity.
2. **Instrument calibration** using certified reference materials.
3. **Contaminant extraction** using acid digestion and solid-phase extraction.
4. **Toxicity testing** with parallel biological endpoints.
5. **Data integration** using statistical modeling and GIS overlays.

Table 6. Technical Configuration Summary

Module	Tools/Software	Function	Frequency
Calibration	Certified CRM standards	Instrument validation	Weekly
QA/QC	Field blanks, duplicates	Quality assurance	Daily

Modeling	PHREEQC, MODFLOW	Transport & reaction modeling	As needed
GIS Integration	ArcGIS, QGIS	Spatial analysis	Biweekly

4. RESULTS AND ANALYSIS

4.1. Geological Composition and Contaminant Distribution

The geological composition of the three selected study sites has a significant impact on the spatial distribution and concentration of contaminants. Site S1 with sulfide-rich bedrock had enhanced levels of heavy metals, lead, and arsenic. Site S2, in an agricultural basin, had high levels of nitrates and phosphate compounds. Mercurial accumulation was revealed in Site S3, a coastal sedimentary region, which is believed to be attributed to saline water interactions (Oberle et al., 2023).

Table 7. Heavy Metal Concentrations Across Geological Sites

Site	Arsenic (mg/kg)	Lead (mg/kg)	Mercury (mg/kg)
S1	45.2	88.1	1.2
S2	12.7	21.3	0.3
S3	28.3	60.5	3.4

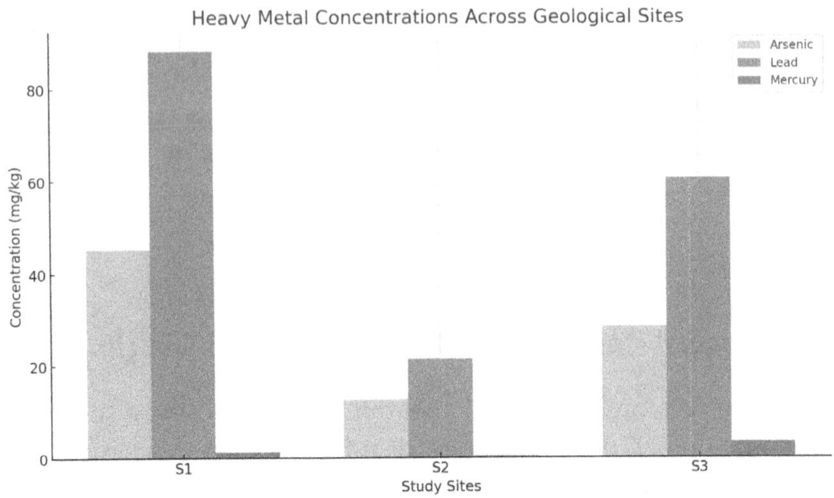

Figure 1. Heavy metal distribution across study sites.

4.2. Bioavailability and Toxicity Trends across Geometrics

Contaminant bioavailability was significantly influenced by matrix type. Percentages of bioavailability were higher for groundwater and sediments than for soils and surface water, especially for arsenic and lead. Mercury has the lowest bioavailability because it is stable in organic matter complexes in clay matrices (Canfield, 2023).

Table 8. Bioavailability of Heavy Metals across Environmental Matrices

Matrix	Arsenic (%)	Lead (%)	Mercury (%)
Soil	35	20	12
Sediment	42	33	22
Groundwater	50	27	18
Surface Water	30	15	10

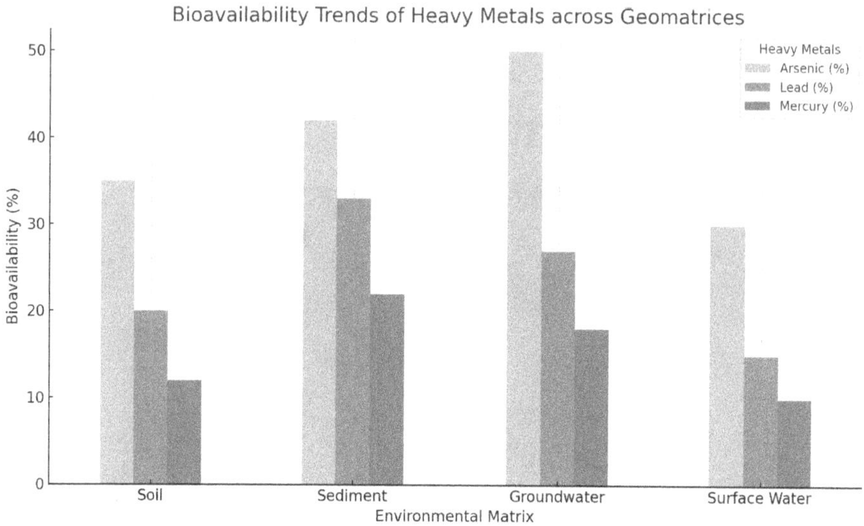

Figure 2. Bioavailability trends across geometrics.

4.3. Correlation of Geochemical Parameters with Ecotoxicity Levels

A statistical correlation analysis revealed that pH had a negative correlation with the toxicity indices, thus suggesting that acidic conditions enhance metal solubility and toxicity. There was a strong inverse relationship between the TDS levels and the organism survival, which means that there was increased stress due to ionic competition (Al-Battashi et al., 2023; Gadd, 2023).

Table 9. Correlation Matrix of Geochemical Parameters and Ecotoxicity

Parameter	pH	EC	TDS	Toxicity Index
pH	1.00	-0.42	0.38	-0.55
EC	-0.42	1.00	-0.62	0.47
TDS	0.38	-0.62	1.00	-0.71
Toxicity Index	-0.55	0.47	-0.71	1.00

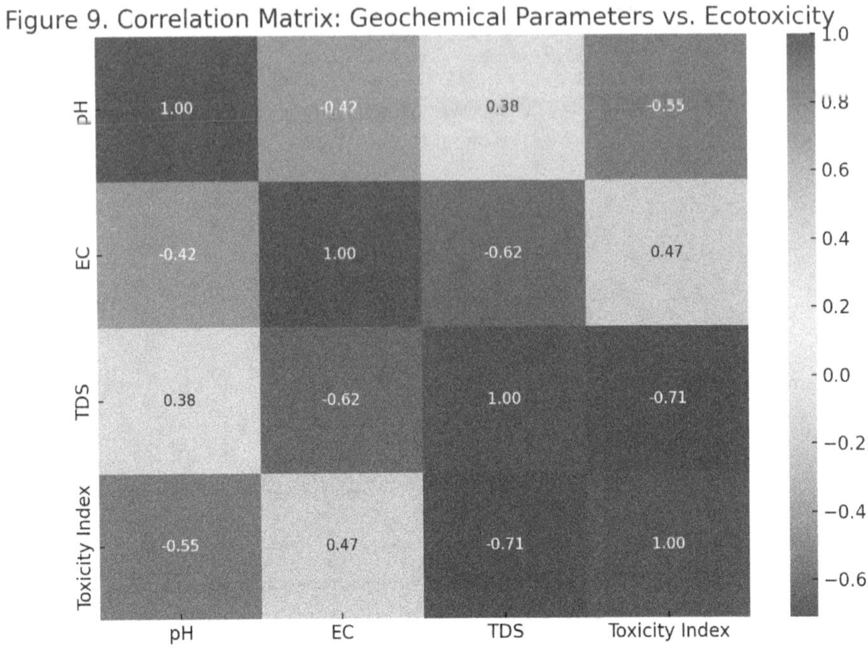

Figure 3. Geochemical and ecotoxicity correlations.

4.4. Spatiotemporal Variations in Contaminant Dynamics

A longitudinal sampling of the aquifers revealed seasonal fluctuations in the arsenic concentrations seasonally, with the highest values recorded in the monsoon season (July–September) when leaching is high and groundwater recharge is enhanced. These findings also highlight the dynamic behavior of contaminants in different climatic cycles (Dellisanti et al., 2023).

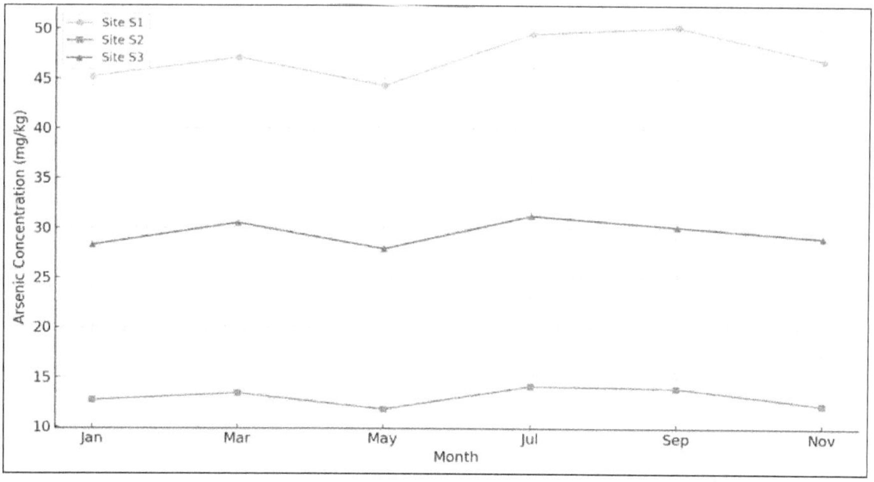

Figure 4. Seasonal variation in arsenic levels across study sites.

4.5. Case Comparisons and Anomalies

Distinct anomalies were observed at each site. Lead levels at Site S1 exceeded WHO limits by over 70%, linked to residual mine tailings. Site S2 showed nitrate spikes, likely due to agricultural runoff. Site S3's mercury levels crossed the ecological threshold, especially during saline intrusion events in the post-monsoon phase (Hudson-Edwards, 2023; Hudson-Edwards et al., 2023a; Lottermoser, 2023).

Table 10. Anomalies Detected Across Sites

Site	Parameter	Observed Level	Threshold Limit	Deviation (%)	Potential Cause
S1	Lead	88.1 mg/kg	50.0 mg/kg	76.2%	Old mine tailings
S2	Nitrate	78.5 mg/L	45.0 mg/L	74.4%	Fertilizer overuse
S3	Mercury	3.4 mg/kg	2.0 mg/kg	70.0%	Saline intrusion

5. DISCUSSION AND RECOMMENDATIONS FOR FUTURE

The results show that the geological condition has a clear effect on the properties of the pollutants, their availability for biota, and their toxicity. For example, the fact that arsenic is more mobile in groundwater in regions with sulfide minerals stresses the role of the geological framework in the transport of metals. Likewise, bottom sediments containing pollutants in coastal areas were found to be more available for biota in saline conditions, which indicates that ion exchange and redox changes are the main factors that control these processes (Canfield, 2023; Hudson-Edwards et al., 2023a). These observations support the idea that excluding the geological factors from ecotoxicological experiments can lead to an underestimation of the environmental impacts, especially in environments with rapidly changing geochemical conditions.

High levels of heavy metals in environmentally important areas not only jeopardize the lives of aquatic and terrestrial organisms but also affect the food production and the well-being of humans via the food chain. Policies based solely on concentration limits do not consider the geological modifiers of pollutants. The application of geo-ecotoxicological assessments in the management of pollution can enhance the current early identification tools and direct resources to areas that require the most attention (Gadd, 2023; Lottermoser, B. G., 2023). This is particularly important in developing countries where the specific local geology increases the chances of exposure.

Some methodological limitations were, however, still present, even though advanced analytical tools were used. For example, standard toxicity indices may not always reflect the complete effects of the geochemical materials on the pollutants. Furthermore, tools like PHREEQC and MODFLOW are theoretical models that are based on some general geological assumptions that may not be accurate for specific sites (Bethke, 2023). There is also a lack of correlation between the biological endpoints and the geospatial information, which hampers the development of real-time ecological risk indicators.

The study recommends that long-term monitoring networks should include geological information, including soil minerals, aquifer porosity, and geochemical fluctuations across the years. Other technologies, including geophysical surveys, mineralogical databases, and sediment histories, can be used to complement conventional pollution sampling to enhance risk assessment. These frameworks can be combined with remote sensing and IoT-based field devices to create a dynamic and location-specific surveillance system (Hudson-Edwards, 2023; Canfield, 2023).

Future should explore:

- Development of integrated modeling platforms combining geochemical kinetics with ecotoxicological responses;
- Machine learning algorithms trained on geo-environmental data for predictive risk mapping;
- Inclusion of microbial-geological interactions in pollutant transformation models;
- Standardization of geo-ecotoxicological indicators in environmental legislation;
- Public participation tools that communicate geological risk in user-friendly formats.

Policy design must evolve to include geoscientific knowledge in land-use planning, particularly in industrial expansion zones and ecological buffer regions.

6. CONCLUSION

This study highlights the indispensable role of geological factors in shaping ecotoxicological dynamics. Key findings include:

- Contaminant mobility and toxicity are highly dependent on lithology and hydrogeology;
- Bioavailability varies significantly across environmental matrices, with sediments and groundwater posing elevated ecological risks;

- Geochemical parameters such as pH, TDS, and EC are strongly correlated with observed toxicity levels;
- Seasonal variations substantially influence contaminant distribution and risk potential.

This chapter contributes to a more holistic environmental assessment model by bridging geological sciences with ecotoxicology. It establishes a scientific basis for incorporating lithological and geochemical attributes into pollution monitoring and risk mitigation frameworks (Rajala et al., 2023; Al-Battashi et al., 2023). The proposed approach supports sustainable land management and reinforces regulatory policies through geological insight while also facilitating targeted remediation strategies. Future efforts must aim for multi-disciplinary collaboration, data transparency, and the development of intelligent, geology-informed environmental decision-making systems. The integration of geological tools and frameworks into ecotoxicology offers a transformative pathway toward sustainable environmental governance; such integration is not only vital for preserving ecosystem integrity but also for ensuring the long-term health and safety of communities dependent on natural resources.

REFERENCES

1. Al-Battashi, H. et al., (2023). The Geomicrobiology of Chromium (VI) Pollution: Microbial Diversity and its Bioremediation Potential. *The Open Biotechnology Journal*, *17*, 1-12.
2. Appelo, C. A. J. & Postma, D. (2023). Geochemistry, Groundwater, and Pollution. Taylor & Francis.
3. Bethke, C. M. (2023). Geochemical and Biogeochemical Reaction Modeling. Cambridge University Press.
4. Canfield, D. E. (2023). Biogeochemistry of Sulfur Isotopes. *Reviews in Mineralogy and Geochemistry*, *89*(1), 559-584.
5. Cheung, K. H. & Gu, J. D. (2023). Mechanism of hexavalent chromium detoxification by microorganisms and bioremediation application

potential: A review. *International Biodeterioration & Biodegradation, 175*, 105312.

6. Choppola, G. et al., (2023). Advances in Agronomy. *Elsevier.*
7. Dellisanti, F. et al., (2023). In-field remediation of tons of heavy metal-rich waste by Joule heating vitrification. *International Journal of Mineral Processing, 169*, 1-8.
8. Gadd, G. M. (2023). Metals, minerals and microbes: geomicrobiology and bioremediation. *Microbiology, 169*(1), 1-14.
9. Hudson-Edwards, K. A. (2023). Natural carbon mineralization and its control on the geochemical evolution of coal-based aquifers in the Salt Range, Punjab, Pakistan. *Environmental Geochemistry and Health, 45*(3), 785-802.
10. Hudson-Edwards, K. A. et al., (2023). Mine Wastes: Past, Present, Future. *Elements, 19*(1), 39-44.
11. Leahy, J. G. & Colwell, R. R. (2023). Microbial degradation of hydrocarbons in the environment. *Microbiological Reviews, 87*(4), 427-448.
12. Liu, J. et al., (2023). A critical review on soil washing during soil remediation for heavy metals and organic pollutants. *International Journal of Environmental Science and Technology, 20*(5), 1-16.
13. Luptakova, A. & Kusnierova, M. (2023). Bioremediation of acid mine drainage contaminated by SRB. *Hydrometallurgy, 150*, 193-197.
14. Ma, L. et al., (2024). Marine exploration and its environmental impact assessment: insights from international standards. *Frontiers in Marine Science, 11*.
15. Merkel, B. J. et al., (2023). Groundwater Geochemistry: A Practical Guide to Modeling of Natural and Contaminated Aquatic Systems. Springer.
16. Oberle, F. K. J. et al., (2023). What a drag: Quantifying the global impact of chronic bottom trawling on continental shelf sediment. *Journal of Marine Systems, 230*, 103634.

17. Oelkers, E. H. & Schott, J. (Eds.). (2023). Thermodynamics and Kinetics of Water-Rock Interaction. *Reviews in Mineralogy and Geochemistry, 70*, 1-569.

18. Palmer, M. A. et al., (2023). Mountaintop Mining Consequences. *Science, 379*(6629), 148-149.

19. Palmer, M. A. et al., (2023). Ecological Restoration of Streams and Rivers: Shifting Strategies and Shifting Goals. *Annual Review of Ecology, Evolution, and Systematics, 54*, 247-269.

20. Rajala, P. et al., (2023). Microbial fouling and corrosion of carbon steel in deep anoxic alkaline groundwater. *Biofouling, 39*(1), 1-15.

21. Rawlings, D. E. (2023). Characteristics and adaptability of iron- and sulfur-oxidizing microorganisms used for the recovery of metals from minerals and their concentrates. *Microbial Cell Factories, 22*(1), 1-7.

22. Smieska, L. et al., (2023). Synchrotron Science for Sustainability: Life Cycle of Metals in the Environment. *arXiv preprint arXiv: 2303*.17502.

23. Zhu, C., & Anderson, G. (2023). Environmental Applications of Geochemical Modeling. Cambridge University Press.

Chapter - 12

HEAVY METAL CONTAMINATION: ADVERSE IMPACT ON THE ENVIRONMENT AND REMEDIATION APPROACHES

Rajat Kumar Gautam[1], Prof. Mukesh Kumar Bhardwaj[1], Sangram Singh[2], Dr Avadhesh Kumar[3] and Dr. Chandra Kant[1]*

[1] Department of Botany, Dharma Samaj College, Aligarh-202001, U.P., India

[2]Department of Botany, Faculty of Life Sciences, Aligarh Muslim University, Aligarh-202002, India

[3]Assistant Professor, Department of Chemistry, Raja Mahendra Pratap Singh University, Aligarh-202001, U.P., India

***Corresponding email: chandrakant@dscaligarh.ac.in**

ABSTRACT

Metal pollution represents a substantial environmental issue resulting from the release of heavy metal ions into the ecosystem. Human activities, such as metal mining, agriculture, and industrial operations, have markedly exacerbated metal pollution in soil, water, and air. Metal contamination poses substantial risks to human health, ecosystem stability, and the global environment. Diverse strategies have been devised to alleviate metal pollution, including physical, biological, and chemical methods. These remediation procedures seek to diminish metal concentrations in the environment, avert more contamination, and rehabilitate degraded ecosystems. This chapter seeks to clarify the origins, attributes, and impacts of heavy metal contamination. Furthermore, we concisely analyse several proven strategies for reducing heavy metal pollution in the environment, along with their advantages and disadvantages.

Keywords: *Environmental pollution, Pollutants, Heavy metals, Toxicology, Anthropogenic activities, Remediation.*

1. INTRODUCTION

The environment comprises the habitat in which humans, flora, fauna, and microorganisms coexist and interact. It has been profoundly tainted by numerous pollutants, such as inorganic ions, organic contaminants, organometallic substances, radioactive isotopes, gaseous pollutants, and nanoparticles (Walker et al., 2012). Environmental pollution refers to an undesirable change in the physical, chemical, or biological characteristics of environmental elements, including air, water, or soil, potentially resulting in harmful effects on various life forms (Gaur et al., 2024).

In recent decades, the rapid advancement of industrialisation and urbanisation, coupled with an increasing human population, has resulted in the deterioration of the quality of vital environmental components-air, water and soil, due to the discharge of hazardous waste from factories, including plastics, heavy metals, nitrates, fossil fuel combustion, acid rain, oil spills, and industrial toxins. These factors pose a considerable threat to biodiversity and the overall ecosystem (Gaur et al., 2024; Prakash and Verma, 2014). Therefore, it is imperative to pursue an urgent resolution for the preservation of modern biodiversity and ecological equilibrium, as both are vital for the shared survival of all life forms, including humans (Kumar, 2017; Verma, 2018; Kolawole and). This chapter explores the sources of heavy metal contamination in the natural environment, the properties of heavy metals, and the possible impacts of specific heavy metals on living organisms. This chapter effectively outlines different remediation strategies, emphasising both their benefits and drawbacks.

2. HEAVY METAL POLLUTION

Heavy metal contamination has emerged as an increasingly serious environmental issue in recent years (Chen et al., 2023), garnering substantial focus because of its toxic aspects, enduring impacts, propensity to accumulate in biological entities, resistance to breakdown, and transference through the food chain (Palansooriya et al., 2022; Wang et al., 2022). Heavy metal contamination poses considerable threats to flora and fauna, while also endangering human well-being and the integrity of ecosystems (Gu et al., 2021).

2.1. Sources of heavy metal pollution

Heavy metals have been present in the Earth's crust since its genesis. The remarkable rise in the utilization of heavy metals has led to an impending growth of metallic substances in both terrestrial and aquatic environments (Briffa et al., 2020). Heavy metals are introduced into the environment from multiple sources, which might include mining, urbanisation, the chemical industry, sewage treatment facilities, pesticide production, biomedical activities, and unsafe agricultural practices (Zaimee et al., 2021; El Ati-Hellal and Hellal, 2021; Briffa et al., 2020) (Fig. 1.1). Natural phenomena may contribute to heavy metal contamination through mechanisms such as eruptions, corrosion of metals, volatilisation of soil and water, sediment resuspension, soil erosion, and geological weathering (Masindi and Muedi, 2018; Walker et al., 2012; Tchounwou et al., 2012).

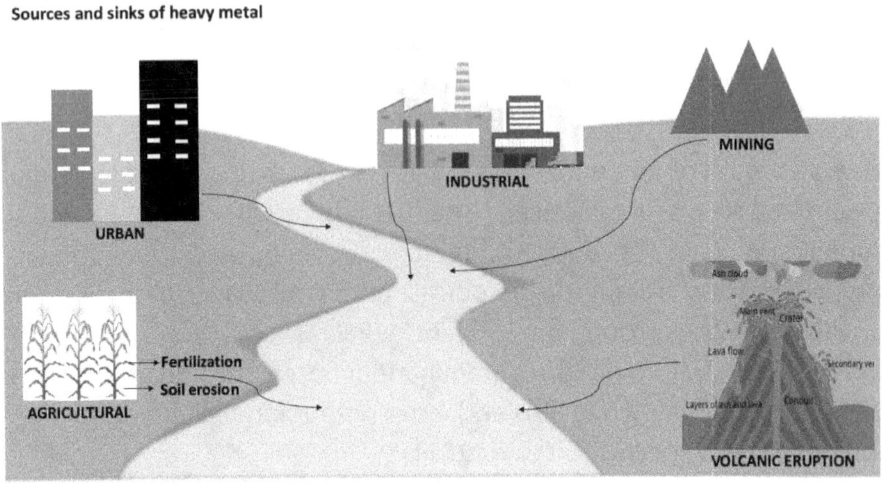

Fig. 1.1 Different sources of heavy metal pollution in the environment

2.2. Characteristics of Heavy Metals

Heavy metals represent a group of naturally occurring, dense metallic elements that exhibit toxicity at low concentrations and have a propensity to bioaccumulate in the natural environment and biological life forms. They are characterized by high atomic weights ranging from 63 to 200, densities exceeding 4000 kg/m^3, low degradability, and the potential to induce long-term health issues (Srivastava and Majumder, 2008; Duffus, 2002). More than 50 elements on the periodic table have been classified as heavy metals, including transition metals, metalloids, lanthanides, and actinides. Included among them are tin, zinc, boron, mercury, copper, nickel, cadmium, cobalt, vanadium, strontium, titanium, arsenic, lead, molybdenum, and chromium (Okpara et al., 2022). Experts have designated approximately 17 of these hazardous materials as highly deadly and relatively accessible. Selected heavy metals, including iron, boron, copper, nickel, and zinc, are essential for plant growth in trace amounts; nevertheless, exceeding the maximum concentration limit (MCL) renders them toxic to both flora and fauna.

3. INFILTRATION AND IMPACT OF HEAVY METALS

Pollutants can permeate the ecosystem via various routes, eventually making their way into the hydrosphere, lithosphere, and atmosphere. Alongside natural processes like volcanic activity and rock weathering, human activities play a substantial role in introducing contaminants into the ecosystem. Such occurrences can arise from unintentional discharges, which encompass shipwrecks, oil spills, mining accidents, and fires; they can also result from the intentional application of biocides, like vector control strategies; in addition, they may stem from waste disposal methods, including industrial effluents and sewage management (Walker et al., 2012).

3.1. Water contamination

Two primary sources contribute to water contamination: urbanization and industrialization. The metals are conveyed through runoff from villages, towns, cities, and industries, ultimately accumulating in the sediments of water bodies (Jeong et al., 2023; Wu et al., 2022a) (Fig.1.1). Even when traces are carried to aquatic environments, they can remain highly toxic to humans and other ecosystems, even at minimal concentrations (Han et al., 2023). The determination of heavy metal toxicity involves several factors, such as the specific metal involved, its classification, the biological function it serves, the organism to which it is exposed, and the duration of that exposure. Negative impacts on one organism have a ripple effect on the entire ecosystem. Given that humans generally occupy the apex of the food chain, the implications for us are significant, as we tend to accumulate higher levels of heavy metals due to the increasing concentration throughout the food chain (He et al., 2023).

3.2. Soil contamination

Soil pollution may arise either through deliberate actions or as a result of unintentional activities. Intentional pollution includes practices such as wastewater irrigation, the use of pesticides, the application of animal manure and fertilisers, lead-based paint, mining ore waste (tailings), sewage sludge, petroleum distillate spills, coal combustion by-products, and trash dumps. The application of untreated sewage and wastewater has led to the build-up of heavy metals in agricultural fields, which are then taken up by crops that humans consume (Fig. 1.1). Unintentional pollution can arise from the flooding of seas and rivers, leading to the introduction of sewage and contaminated water onto land, alongside incidents involving trucks transporting hazardous materials. Heavy metals remain in the soil for long durations because they do not degrade and cannot be broken down through microbiological or chemical processes. Soils contaminated with heavy metals hinder the growth and survival of plants, leading to significant nutritional, ecological, and evolutionary challenges (Mani and Kumar, 2014). These alterations can lead to lipid oxidation and protein damage, which in turn can impact the morphology, metabolism, and growth of microorganisms (Xie et al., 2016). Heavy metals influence the diversity, abundance, and functionality of microbial communities, along with the genetic composition of microorganisms. Heavy metal toxicity has the potential to modify the structure of nucleic acids, disrupt cell membranes, interfere with cellular functions, inhibit enzyme activity, and impact energy production. Such modifications can result in lipid oxidation and protein damage, ultimately affecting the morphology, metabolism, and growth of microorganisms (Xie et al., 2016).

3.3. Air contamination

Similar to water contamination, air pollution has resulted from urbanization and industrialization. Pollutants infiltrate the atmosphere in various forms. They may enter as particles, droplets, gaseous forms, or in combination with particles or droplets. Particles and droplets typically do not traverse extensive distances and generally descend to the earth after a brief interval, but if they are diminutive in size, they may cover greater distances. Particles in the gaseous form can be conveyed over extensive distances by air masses. Natural and anthropogenic activities have resulted in the emission of particulate matter (PM), particularly fine particles and dust. Particulate matter originating from natural processes is emitted during sandstorms, volcanic eruptions, soil erosion, and rock weathering. Particulate matter resulting from human activity is emitted through industrial processes, fossil fuel combustion, vehicle exhaust, smelting, and other sources. Particulate matter can precipitate significant health issues, contribute to infrastructure degradation, induce acid rain, cause corrosion, lead to eutrophication via deposition in water during precipitation, and result in haze formation (Briffa et al., 2020).

Additional causes of atmospheric pollution include internal combustion engines and jet engines. Catalytic converters and unleaded gasoline have contributed to the reduction of vehicular pollution, in addition to enhancing engine performance. Diesel engines, vintage automobiles, and excessive vehicle numbers continue to be an issue. The use of pesticides constitutes an additional source of pollutants alongside refrigerators, aerosols, and radioactive contamination (Walker et al., 2012). Air contaminated with heavy metals poses serious risks to both plant and human health. This occurs mainly through disruptions in cellular functions due to direct exposure and the accumulation of these metals in the food chain when plants take them up from the air. Consequently, this can lead to various health issues upon human consumption. Notable effects include stunted growth in plants, impaired photosynthesis, and in humans, respiratory problems, organ damage, and developmental issues, which vary depending on the specific heavy metal present (Mitra et al., 2022) **(Fig.1.2).**

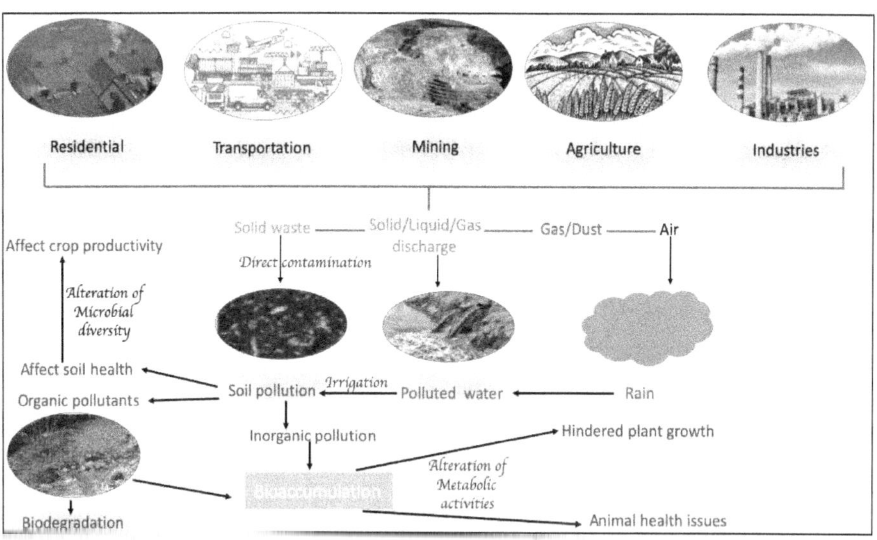

Figure. 1.2. Impact of heavy metals on the environment

4. CONTROL AND REMEDIATION APPROACHES

The detrimental impact of heavy metals on the environment, attributed to their toxicity, non-biodegradability, and persistence, is a global issue. Consequently, their elimination from the environment has garnered the attention of several industrial and research professionals. A variety of approaches and technologies have been devised to eradicate or diminish heavy metals in the environment. Remediation approaches can be classified into three categories: physical, chemical, and biological. The selection of a treatment method for managing heavy metals in a specific context is contingent upon various aspects, including the characteristics of the heavy metals, the permissible limits, and the concentrations of influent and effluent. The applicability of the various approaches is contingent upon removal efficiency, environmental impact, and economic viability (Vareda et al., 2019). The aforementioned methods have been succinctly addressed in this section.

4.1. PHYSICAL REMEDIATION APPROACHES

4.1.1. Chemical Coagulation and Flocculation

Chemical coagulation and flocculation can be utilized to eliminate hazardous heavy metals and suspended particulates. Chemical coagulants are introduced to the wastewater solution, causing the destabilization of colloidal particles into aggregates (Fu and Wang, 2011). Flocculation is often used subsequent to coagulation to aggregate destabilized particles into larger entities. Thereafter, sedimentation and filtration are used to separate these bigger particles. A diverse array of treatment chemicals exists for wastewater management, including coagulants like ferrous sulfate, ferric chloride, and aluminum sulfate, as well as flocculants such as polyelectrolytes, including polyferric sulfate, polyacrylamide, and polyaluminum chloride (Vareda et al., 2019; Tang et al., 2016). The efficacy of coagulation is influenced by factors like the kind of coagulant, coagulant concentration, mixing conditions, temperature, alkalinity, and pH (Carolin et al., 2017; Bratby, 2016). Although coagulation and flocculation are efficacious ways for the removal of heavy metals, it is frequently essential to augment these techniques with additional approaches to attain comprehensive removal of heavy metals. These approaches also result in the production of by-products (flocs) that may function as secondary pollutants.

4.1.2. Adsorption

The adsorption method is considered the most economical, efficient, and selective remediation technology for the management of heavy metals in wastewater (Sharma et al., 2019; Sun et al., 2019; Da'na, 2017). The preparation process is uncomplicated, and it effectively enables the complete elimination of heavy metals. This process entails a solid-liquid mass transfer mechanism where the heavy metal (adsorbate) migrates from the wastewater to the solid interface (adsorbent) and subsequently adheres via physisorption or chemisorption. Physisorption is a process that occurs due to weak van der Waals forces, while chemisorption involves the formation of strong covalent bonds between the adsorbate and the adsorbent. Activated carbon serves as the most commonly used adsorbent for the elimination of hazardous heavy metals. The material's enhanced affinity for heavy metals, combined with its increased surface area, makes it a preferred choice among current adsorbents. However, its application is constrained due to its restricted regenerative capacity and elevated expense; consequently, biomaterials derived from agricultural and industrial activities have been extensively utilized as alternative adsorbents for heavy metal treatment (Vardhan et al., 2919).

4.1.3. Electrochemical Methods

Electrochemical methods are recognized as an effective approach for achieving high removal efficiency of heavy metals in industrial wastewater. Typically, these approaches are meticulously regulated, produce minimal sludge, necessitate fewer chemicals, and exhibit quicker kinetics in the removal of heavy metals from industrial wastewater. Nonetheless, they frequently come with significant upfront costs and elevated energy consumption expenses, which can hinder their broader implementation. Electrochemical methods involve utilizing cathodic and anodic reactions to eliminate heavy metals in their elemental forms (Pulkka et al., 2014). Common electrochemical methods for the removal of metals encompass electrodeposition, electro-floatation, electrocoagulation, electro-flocculation, electro-oxidation, electrochemical oxidation, and electrochemical reduction (Chen et al., 2023; Mezine et al., 2018; Adjeroud et al., 2018; Chen et al., 2013).

4.1.4. Ion Exchange

This technique is commonly employed for the removal of metals from the environment due to its rapid kinetics, high efficiency in removal, minimal or negligible sludge production, and enhanced treatment capabilities. The process of ion substitution in the solution occurs with similar ions present in the insoluble substance. The substance is referred to as ion-exchange resin (Maletskyi, 2020; Edebali and Pehlivan, 2016). The process entails channelling wastewater through the ion-exchange column to the bed, effectively removing the heavy metals contained in the effluent. In cases of elevated heavy metal concentration within the column, backwashing is conducted to eliminate the deposited heavy metals. Synthetic and natural ion exchange resins can be classified into two categories. Research shows that synthetic resin materials are preferred for their effectiveness in removing heavy metals. The ion exchange process has a notable limitation, which is the generation of harmful secondary pollutants resulting from the chemicals commonly employed for regeneration. The procedure involves considerable costs attributed to the necessity for large quantities of resin for wastewater treatment (Masindi and Muedi, 2018; Carolin et al., 2017).

4.1.5. Membrane Filtration

This technology is mostly regarded and employed for the treatment of inorganic effluents. The technique can eliminate suspended materials and both organic and inorganic constituents, including heavy metals. Various membrane filtration techniques, including nanofiltration, reverse osmosis, and ultrafiltration, can be employed based on particle size to remove heavy metals from wastewater. The ultrafiltration technology enables the filtration of heavy metals, suspended particles, and macromolecules, contingent upon pore size (5-20 micrometres) and the molecular weight of the separated compounds (1000-100,000 Da).

Ultrafiltration can eliminate over 90% of metal concentrations between 10 and 112 mg/L at a pH range of 5 to 9.5 under a pressure of 2.5 bar. Ultrafiltration has several advantages, including high packing density, reduced driving force, and minimal spatial requirements (Yadav et al., 2019).

4.2. CHEMICAL REMEDIATION APPROACHES

4.2.1. Chemical Precipitation

This approach is frequently implemented for the extraction of harmful heavy metals from industrial wastewater due to its economic efficiency and ease of implementation (Izadi et al., 2017). In the process of chemical precipitation, a precipitating agent is introduced into wastewater that contains heavy metal ions, resulting in the formation of insoluble precipitates. This approach is recorded to demonstrate considerable efficacy in the removal of Cu, Mn, Zn, and Cd metals (Carolin et al., 2017). Additional methods, such as sedimentation and filtration, are often necessary to eliminate precipitates from processed water (Fu and Wang, 2011). Prominent precipitation processes include hydroxide precipitation, which utilises agents such as calcium hydroxide, lime, and sodium hydroxide to produce hydroxide precipitates, as well as sulphide precipitation, a proficient technique that employs gaseous (H_2S), aqueous (NH_4S, Na_2S, NaHS), and solid (CaS, FeS) forms (Vardhan et al., 2019; Carolin et al., 2017).

4.3. BIOREMEDIATION OR BIOLOGICAL APPROACHES

The constraints of various remediation procedures, including elevated costs, toxic waste production, lack of specificity in metal-binding capabilities, and substantial reagent demands, urge the exploration of alternative approaches like bioremediation (Ayangbenro and Babalola, 2017). This methodology is an innovative approach for the extraction of hazardous materials from aquatic ecosystems through the use of plants, animals, and biological systems, including microorganisms (Head, 1998). This approach has displayed effectiveness in removing heavy metals from polluted soils and wastewater, supporting sustainable practices in the rehabilitation of contaminated systems and restoring the natural state of the adversely affected environment, yielding enduring ecological advantages at minimal expense (Dixit et al., 2015). Bioremediation strategies are categorized into in situ (on-site) and ex-situ approaches. Common approaches include phytoremediation (utilizing plants), bio-stimulation (enhancing natural microbial activity), bioaugmentation (introducing beneficial microbes), as well as bio-venting and bio-slurping for the remediation of soil and groundwater.

4.3.1. Phytoremediation

Phytoremediation utilizes plants to remove, stabilize, or transform pollutants from soil, water, and air through eco-friendly mechanisms. Key techniques include:

- **Phytoextraction (Phytoaccumulation):** This involves the uptake of heavy metals from soil and their transport to aerial parts of plants, where metals are stored in vacuoles. Metal phytochelatin complexes facilitate absorption and translocation. Hyperaccumulator species can accumulate excessive concentrations of metals such as arsenic and lead (; Gul et al., 2021; Asgari Lajayer et al., 2019).

- **Rhizofiltration:** Roots extract pollutants from wastewater, groundwater, or surface water rather than soil. Aquatic and semi-aquatic plants, including sunflower, tobacco, and spinach, are effective in removing heavy metals such as cadmium, lead, and chromium, which primarily accumulate in the roots (Mohan et al., 2021).

- **Phytostabilization:** Plant roots limit the mobility of heavy metals, reducing their environmental impact. Pollutants adhere to root surfaces via transport proteins or secondary metabolites and may be broken down into simpler forms through interaction with proteins, amino acids, or sugar derivatives. Species like black nightshade, sunflower, and cowpea are commonly used for this purpose (Li et al., 2020).

- **Phytovolatilization:** Plants take up contaminants from soil and convert them into less toxic volatile compounds that are released into the atmosphere through leaves and stems. This method is effective for contaminants of lower toxicity (Pouresmaieli et al., 2022).
- **Phytodegradation (Phytotransformation):** Plants metabolize or degrade organic pollutants into non-toxic compounds within their tissues before distributing them to other plant organs. This approach is particularly effective for detoxifying organic contaminants (Nebeská et al., 2021).

These phytoremediation strategies are cost-effective, environmentally sustainable, and serve as practical solutions for mitigating pollution in soil and water systems.

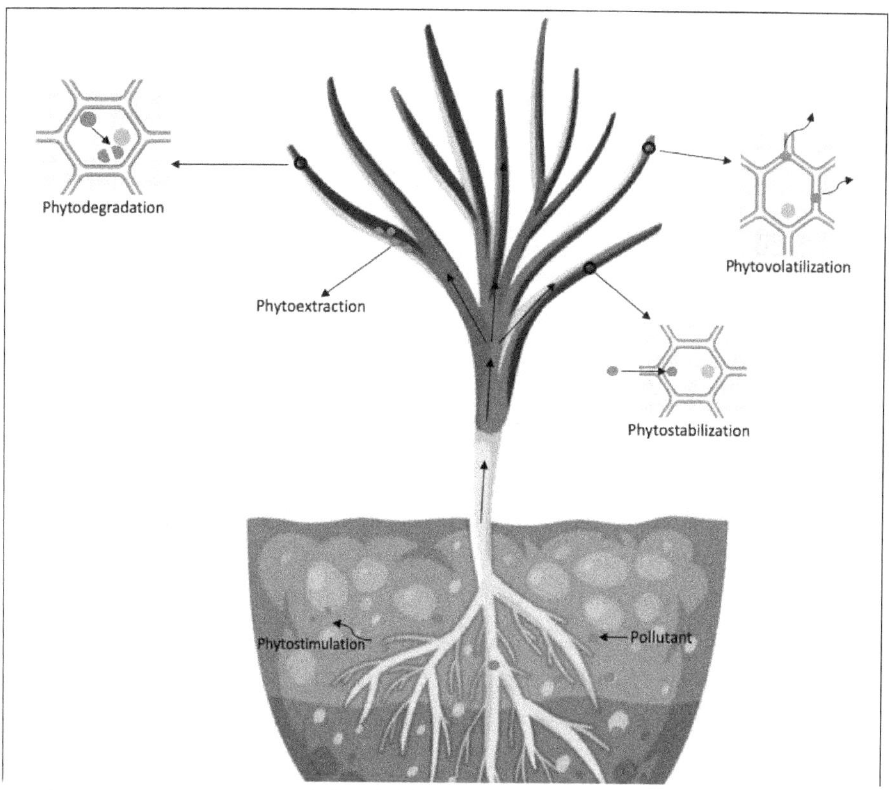

Figure.1.3. Different types of phytoremediation of heavy metals

4.3.2. Microbial-assisted remediation of heavy metals

Microbial remediation employs living microorganisms, such as bacteria, fungi, and archaea, to decompose and neutralise various chemical and metallic hazardous wastes in the environment (Yaashikaa et al., 2022). Bioremediation involves the direct use of microorganisms at contaminated sites to facilitate the degradation of pollutants. Microorganisms are essential in numerous remediation methods, including bioaugmentation and bio-stimulation. Bioaugmentation involves the introduction of specific microorganisms to a contaminated site to improve pollutant degradation. Bio-stimulation entails modifying environmental conditions at a location to promote the growth and activity of indigenous microorganisms that can degrade contaminants. Bioremediation technologies have significantly advanced, establishing them as an effective method for mitigating soil and water contamination. Bioremediation of oil spills is the most economically viable and environmentally sustainable method among the various techniques available (Pete et al., 2021).

4.3.3. Role of genetic engineering microbes in bioremediation

Genetically engineered organisms represent a promising strategy for advancing bioremediation initiatives (Irshad et al., 2021; Tonelli et al., 2021). Additional investigation is necessary to enhance bioremediation methodologies in engineering that can proficiently eradicate the intricate combinations of pollutants present in various ecosystems. Microorganisms, including bacteria and fungi, engage in complex metabolic processes that utilise pollutants as energy sources, thereby playing an essential role in the degradation and detoxification of harmful substances. Contemporary scientific endeavours leverage biotechnology to advance the creation of genetically modified organisms (GMOs), which are essential for facilitating comparisons with their wild-type counterparts. Genetically modified organisms possess crucial protein machinery that allows them to absorb and regulate heavy metals through the use of gene regulatory elements such as promoters, binders, and terminators. A multitude of specialists contend that certain bacteria possess the capability to remove heavy metals from their environment, either through accumulation or by developing resistance to these elements. A diverse array of microorganisms, including *Acinetobacter sp., Alcaligenes odorans, Aspergillus niger (fungus), Aspergillus versicolour, Bacillus subtilis, Corynebacterium propinquum, Fomitopsis pinicola, Microbacterium sp., Pseudomonas sp., P. putida, P. aeruginosa, Ralstonia sp., and Streptomyces,* play a significant role in the remediation of heavy metals. (Irankhah et al., 2019; Bahadure et al., 2013; Wasi et al., 2013; Abdulsalam et al., 2012).

CONCLUSIONS

Heavy metal toxicity is recognised as one of the most severe environmental concerns because it causes reduced crop productivity and directly or indirectly affects the survival of practically all kinds of life on Earth. The present chapter summarizes physical, chemical, and biological strategies for heavy metal elimination, encompassing both metabolically dependent and independent approaches. Additionally, it addresses the sources of heavy metals and their associated toxicity. This study enhances our comprehension of designing effective biological wastewater treatment systems.

REFERENCES

1. Abdulsalam, S., Adefila, S., Bugaje, I., Ibrahim, S.(2012). Bioremediation of soil contaminated with used motor oil in a closed system. *J Bioremediat Biodegrad, 03*,1-7.

2. Adjeroud, N., Elabbas, S., Merzouk, B., Hammoui, Y., Felkai-Haddache, L., Remini, H., Leclerc, J., Madani, K. (2018). Effect of Opuntia ficus indica Mucilage on Copper Removal from Water by Electrocoagulation-Electroflotation Technique. *J. Electroanal. Chem, 811*, 26–36.

3. Asgari Lajayer, B., Khadem Moghadam, N., Maghsoodi, M.R., Ghorbanpour, M., Kariman, K. (2019). Phytoextraction of heavy metals from contaminated soil, water, and atmosphere using ornamental plants: mechanisms and efficiency improvement strategies. *Environ Sci Pollut Res, 26*, 8468-8484.

4. Ayangbenro, A. S., Babalola, O. O. (2017). A New Strategy for Heavy Metal Polluted Environments: A Review of Microbial Biosorbents. *Int. J. Environ. Res. Public Health, 14* (1), 94.

5. Bahadure, S., Kalia, R., Chavan, R. (2013). Comparative study of bioremediation of hydrocarbon fuels. *Int J Biotechnol Bioeng Res, 4,* 677-686.

6. Bratby, J. (2016). Coagulation and Flocculation in Water and Wastewater Treatment; IWA Publishing: (3rd ed., Vol. *15*).

7. Briffa, J., Sinagra, E., Blundell, R. (2020). Heavy metal pollution in the environment and its toxicological effects on humans. Heliyon, *6*(9):e04691.

8. Carolin, C. F., Kumar, P. S., Saravanan, A., Joshiba, G. J., Naushad, M. (2017). Efficient Techniques for the Removal of Toxic Heavy Metals from Aquatic Environment: A Review. *J. Environ. Chem. Eng, 5* (3), 2782–2799.

9. Chen, D., Wang, X. H., Luo, X. M., Huang, G. X., Tian, Z., Li, W. Y. (2023). Delineating and identifying risk zones of soil heavy metal pollution in an industrialized region using machine learning. *Environ. pollut. (Amsterdam Neth.), 318,* 120932.

10. Chen, W., He, X., Jiang, Z., Li, B., Li, X.-y., Lin, L. (2023). A Capacitive Deionization and Electro-oxidation Hybrid System for Simultaneous Removal of Heavy Metals and Organics from Wastewater. *Chem. Eng. J, 451,* 139071.

11. Chen, X., Huang, G., Wang, J. (2013). Electrochemical Reduction/ Oxidation in the Treatment of Heavy Metal Wastewater. *J. Metall. Eng. (ME), 2* (4), 161.

12. Da'na, E. (2017). Adsorption of Heavy Metals on Functionalized-Mesoporous Silica: A Review. *Micropor. Mesopor. Mater, 247,* 145–157.

13. Das, M., Bramhanand, P.S., Laxminarayana, K. (2021). Performance and efficiency services for the removal of hexavalent chromium

from water by common macrophytes. *Int J Phytoremediation, 23,* 1095-1103.

14. Dixit, R., Malaviya, D., Pandiyan, K., Singh, U., Sahu, A., Shukla, R., Singh, B., Rai, J., Sharma, P., Lade, H., Paul, D. (2015). Bioremediation of Heavy Metals from Soil and Aquatic Environment: An Overview of Principles and Criteria of Fundamental Processes. *Sustainability, 7* (2), 2189–2212.

15. Duffus, J. H. "Heavy metals" — a meaningless term? (IUPAC Technical Report). *Pure and Applied Chemistry, 74* (5), 793–807.

16. Edebali, S., Pehlivan, E. (2016). Evaluation of Chelate and Cation Exchange Resins to Remove Copper Ions. *Powder Technol, 301,* 520–525

17. El Ati-Hellal, M., Hellal, F. (2021). Heavy Metals in the Environment and Health Impact. Intech Open. https://doi.org/10.5772/intechopen.97204

18. Fu, F., Wang, Q. (2011). Removal of Heavy Metal Ions from Wastewaters: A Review. *J. Environ. Manag, 92* (3), 407–418.

19. Gaur, N., Sharma, S., Yadav, N. (2024). Chapter 2 - Environmental pollution, Editor(s): Vinod Kumar Garg, Anoop Yadav, Chandra Mohan, Sushma Yadav, Neeraj Kumari, In Advances in Green and Sustainable Chemistry, Green Chemistry Approaches to Environmental Sustainability, *Elsevier*, 23-41,

20. Gu, Y. G., Wang, Y. S., Ouyang, J., Jordan, R. W., and Jiang, S. J. (2021). Impacts of coastal aquaculture on sedimentary phosphorus speciation and fate: evidence from a seaweed cultivation area off Nan'ao Island, South China. *Mar. pollut. Bull, 171*, 112719.

21. Gul, I., Manzoor, M., Hashim, N., et al. (2021). Challenges in microbially and chelate-assisted phytoextraction of cadmium and lead – a review. *Environ Pollut, 287,*117667.

22. Han, Y., Xun, F., Zhao, C., Li, B., Luo, W., Feng, M., Xu, D., Xing, P., Wu, Q.L. (2023). Evaluating potential ecological risks of emerging toxic

elements in lacustrine sediments: A case study in Lake Fuxian, China, *Environmental Pollution, 323*, 121277

23. He, L., Wang, S., Liu, M., Chen, Z., Xu, J., Dong, Y. (2023). Transport and transformation of atmospheric metals in ecosystems: A review, *Journal of Hazardous Materials Advances, 9*, 100218.

24. Head, I. M. (1998). Bioremediation: Towards a Credible Technology. *Microbiology (Reading),144* (3), 599–608.

25. Huang, D., Xiao, R., Du, L. (2021). Phytoremediation of poly- and perfluoroalkyl substances: a review on aquatic plants, influencing factors, and phytotoxicity. *J Hazard Mater, 418,* 126314.

26. Irankhah, S., Abdi Ali, A., Mallavarapu, M. (2019). Ecological role of Acinetobacter calcoaceticus GSN3 in natural biofilm formation and its advantages in bioremediation. *Biofouling, 35*, 377-391.

27. Irshad, S., Xie, Z., Mehmood, S., Nawaz, A., Ditta, A., Mahmood, Q.(2021). Insights into conventional and recent technologies for arsenic bioremediation: a systematic review. *Environ Sci Pollut Res. 28,* 18870-18892.

28. Izadi, A., Mohebbi, A., Amiri, M., Izadi, N. (2017). Removal of Iron Ions from Industrial Copper Raffinate and Electrowinning Electrolyte Solutions by Chemical Precipitation and Ion Exchange. *Miner. Eng, 113*, 23–35.

29. Jeong, E., Kim, Y.N., Jin-Yong Lee, J.Y., Raza, M. (2023). Microplastic contamination in groundwater of rural area, eastern part of Korea, Science of The Total Environment, 895, 165006

30. Kolawole, A.S., Iyiola, A.O. (2023). Environmental Pollution: Threats, Impact on Biodiversity, and Protection Strategies. In: Izah, S.C., Ogwu, M.C. (eds) Sustainable Utilization and Conservation of Africa's Biological Resources and Environment. *Sustainable Development and Biodiversity, 32,* Springer, Singapore.

31. Kumar, A.V. (2017). Necessity of Ecological Balance for Widespread Biodiversity. *Indian Journal of Biology*, *4*(2), 158-160.

32. Kumar, S., Thakur, N., Singh, A.K., Gudade, B.A., Ghimire, D., Das, S. (2022). Aquatic macrophytes for environmental pollution control: phytoremediation technology for the Removal of Heavy Metals and Other Contaminants From Soil and Water. *Elsevier,* 291-308.

33. Li, J., Zheng, B., Chen, X., Li, Z., Xia, Q., Wang, H., Yang, Y., Zhou, Y., & Yang, H. (2021). The Use of Constructed Wetland for Mitigating Nitrogen and Phosphorus from Agricultural Runoff: A Review. *Water, 13*(4), 476. https://doi.org/10.3390/w13040476

34. Li, X., Zhu, W., Meng, G., Guo, R., Wang, Y. (2020). Phytoremediation of alkaline soils co-contaminated with cadmium and tetracycline antibiotics using the ornamental hyperaccumulators *Mirabilis jalapa* L. and *Tagetes patula* L. *Environ Sci Pollut Res, 27,* 14175-14183.

35. Liu, Z., Tran, K.Q. (2021). A review on the disposal and utilization of phytoremediation plants containing heavy metals. *Ecotoxicol Environ Saf,* 226112821.

36. Maletskyi, Z. (2020). Advances in Membrane Materials and Processes for Water and Wastewater Treatment. In *Multidisciplinary Advances in Efficient Separation Processes; American Chemical Society,* 3–35.

37. Mani, D., Kumar, C. (2014). *Int. J. Environ. Sci. Technol, 11,* 843–872.

38. Masindi, V., Muedi, K. L. (2018). Environmental Contamination by Heavy Metals. *Heavy Met, 10,* 115–132.

39. Mezine, Z., Kadri, A., Hamadou, L., Benbrahim, N., Chaouchi, A. (2018). Electrodeposition of Copper Oxides (CuxOy) from Acetate Bath. *J. Electroanal. Chem, 817,* 36–47; DOI: 10.1016/j.jelechem.2018.03.055.

40. Mitra, S., Chakraborty, A.J., Tareq, A.M., Emran, T.B., Nainu, F., Khusro, A., Idris, A.M., Khandaker, M.U., Osman, H., Alhumaydhi, F.A., Simal-Gandara, J. (2022). Impact of heavy metals on the environment and human health: Novel therapeutic insights to counter the toxicity. *Journal of King Saud University - Science, 34* (3),101865.

41. Mohan, I., Goria, K., Dhar, S., Kothari, R., Bhau, B., Pathania, D. (2021). Phytoremediation of heavy metals from the biosphere perspective

and solutions. *Pollutants and Water Management: Resources, Strategies and Scarcity, 16*, 95-127.

42. Nebeská, D., Trögl, J., Ševců, A. (2021). Miscanthus x giganteus in phytodegradation and changes in the bacterial community of soil contaminated by the petroleum industry. *Ecotoxicol Environ Safety. 224*,112630.

43. Okpara, E. C., Fayemi, O. E., Wojuola, O. B., Onwudiwe, D. C., Ebenso, E. E. (2022). Electrochemical Detection of Selected Heavy Metals in Water: A Case Study of African Experiences. *R.S.C. Adv, 12* (40), 26319–26361.

44. Palansooriya, K. N., Li, J., Dissanayake, P. D., Suvarna, M., Li, L., Yuan, X. (2022). Prediction of soil heavy metal immobilization by biochar using machine learning. *Environ. Sci. Technol, 56*, 4187–4198.

45. Pete, A.J., Bharti, B., Benton, M.G. (2021). Nano-enhanced bioremediation for oil spills: a review. *ACS ES & T Eng. 1*, 928-946.

46. Pouresmaieli, M., Ataei, M., Forouzandeh, P., Azizollahi, P., Mahmoudifard, M. (2022). Recent progress on sustainable phytoremediation of heavy metals from soil. *J Environ Chem Eng, 10,*108482.

47. Prakash, S., Verma, A.K. (2014). Effect of Organophosphorus pesticide (Chlorpyrifos) on the haematology of Heteropneustes fossilis (Bloch). *International Journal of Fauna and Biological Studies, 1*(5), 95-98.

48. Prasad, M.N.V. (2019). Prospects for manipulation of molecular mechanisms and transgenic approaches in aquatic macrophytes for remediation of toxic metals and metalloids in wastewaters. *Transgenic Plant Technology for Remediation of Toxic Metals and Metalloids. Elsevier,* 395-428.

49. Pulkka, S., Martikainen, M., Bhatnagar, A., Sillanpää, M. (2014). Electrochemical Methods for the Removal of Anionic Contaminants from Water–A Review. *Sep. Purif. Technol, 132*, 252–271.

50. Sharma P. (2021). Efficiency of bacteria and bacterial-assisted phytoremediation of heavy metals: an update. *Bioresour Technol, 328,* 124835.

51. Sharma, M., Singh, J., Hazra, S., Basu, S. (2019). Adsorption of Heavy Metal Ions by Mesoporous ZnO and TiO2@ZnO Monoliths: Adsorption and Kinetic Studies. *Microchem. J, 145,* 105–112.

52. Shweta, Y. (2019). On the Physical Chemical and Bioremediation Methods for the Removal of Certain Heavy Metals. *Bioscience Biotechnology Research Communications. 12.* 1063-1071.

53. Srivastava, N. K., Majumder, C. B. (2008). Novel Biofiltration Methods for the Treatment of Heavy Metals from Industrial Wastewater. *J. Hazard. Mater, 151* (1), 1–8.

54. Sun, S., Zhu, J., Zheng, Z., Li, J., Gan, M. (2019). Biosynthesis of β-Cyclodextrin Modified Schwertmannite and the Application in Heavy Metals Adsorption. *Powder Technol, 342,* 181–192.

55. Tang, X., Zheng, H., Teng, H., Sun, Y., Guo, J., Xie, W., Yang, Q., Chen, W. (2016). Chemical Coagulation Process for the Removal of Heavy Metals from Water: A Review. *Desalin. Water Treat, 57* (4), 1733–1748.

56. Tchounwou, P.B., Yedjou, C.G., Patlolla, A.K., Sutton, D.J. (2012). Heavy metal toxicity and the environment. *Exp Suppl,* 101, 133-64.

57. Tonelli, F.M.P., Lemos, M.S., Tonelli, F.C.P.(2021). *Genetically Modified Organisms as tools for water bioremediation, freshwater pollution, and aquatic ecosystems.* Apple Academic Press, 301-320.

58. Tripathy, P.K., Mohapatra, M., Pattnaik, R., Tarafdar, L., Panda, S., Rastogi, G. (2022). Macrophyte Diversity and Distribution in Brackish Coastal Lagoons: A Field Survey From Chilika, Odisha Coastal Ecosystems. *Springer,* 325-358.

59. Vardhan, K. H., Kumar, P. S., Panda, R. C. (2019). A Review on Heavy Metal Pollution, Toxicity and Remedial Measures: Current Trends and Future Perspectives. *J. Mol. Liq,* 290, 111197

60. Vareda, J. P., Valente, A. J. M., Durães, L. (2019). Assessment of Heavy Metal Pollution from Anthropogenic Activities and Remediation Strategies: A Review. *J. Environ. Manag, 246*, 101–118.

61. Verma, A.K. (2018). Ecological Balance: An Indispensable Need for Human Survival. *Journal of Experimental Zoology India*, *21*(1), 407-409.

62. Walker, C.H., Sibly, R.M., Hopkin, S.P., Peakall, D.B. (2012). In: Principles of Ecotoxicology; Group, T. and F., (4th ed.). CRC Press. https://doi.org/10.1201/b11767

63. Wang, X. N., Gu, Y. G., and Wang, Z. H. (2022). Rare earth elements in different trophic level marine wild fish species. *Environ. pollut. (Amsterdam Neth.), 292*, 118346.

64. Wasi, S., Tabrez, S., Ahmad, M. (2013). Use of Pseudomonas spp. for the bioremediation of environmental pollutants: a review. *Environ Monit Assess, 185,* 8147-8155.

65. Wu, C., Xiong, X., Hamidian, A.H., Zhang, Y., Xu, X. (2022). A review on source, occurrence, and impacts of microplastics in freshwater aquaculture systems in China. *Water Biology and Security, 1* (3), 100040

66. Xie, Y., Fan, J., Zhu, W., Amombo, E., Lou, Y., Chen, L., Fu, J.(2016). Effect of Heavy Metals Pollution on Soil Microbial Diversity and Bermudagrass Genetic Variation. Front Plant Sci, *7*,755.

67. Yaashikaa, P.R., Kumar, P.S., Jeevanantham, S., Saravanan, R. (2022). A review on bioremediation approach for heavy metal detoxification and accumulation in plants. *Environ Pollut, 301*,119035.

68. Zaimee, M.Z.A., Sarjadi, M.S., Rahman, M.L. (2021). Heavy Metals Removal from Water by Efficient Adsorbents. *Water*, *13* (19), 2659.

Chapter - 13

SUSTAINABLE POLLUTION CONTROL: GREEN CHEMISTRY, BIOREMEDIATION & WASTE MANAGEMENT

Dr. Renu Kumari[1*] and Dr. Manju Bhaskar[2]

[1]Department of Chemistry, M. M. H. College, Ghaziabad 201009 Uttar Pradesh, India

[2]Department of Zoology, M. M. H. College, Ghaziabad 201009 Uttar Pradesh, India

Corresponding email: renummh@gmail.com

ABSTRACT

Green chemistry (sometimes called sustainable chemistry) is the branch of chemistry that deals with the design and optimization of processes and products to lower, or remove altogether, the production and use of toxic substances. Green chemistry is not the same as environmental chemistry. As environmental chemistry focuses on the effects of polluting chemicals on nature, green chemistry focuses on the environmental impact of chemistry, including lowering consumption of nonrenewable resources and technological approaches for controlling environmental pollution (Sheldon et al., 2007 and Clark, 2012).

1. KEY PRINCIPLES OF GREEN CHEMISTRY

The twelve principles put forward by the American chemists Paul Anastas and John Warner In the year 1998 to lay the foundation for green chemistry are listed below.

- **Prevention of waste:** Preventing the formation of waste products is always preferable to the clean-up of the waste once it is generated.
- **Atom economy:** The synthetic processes and methods that are based on green chemistry must always try to maximize the consumption and incorporation of all the raw materials into the final product. This must strictly be followed to minimize the waste generated by any process.
- **Avoiding the generation of hazardous chemicals:** Reactions and processes that involve the synthesis of certain toxic substances that pose hazards to human health must be optimized in order to prevent the generation of such substances.
- **The design of safe chemicals:** During the design of chemical products that accomplish a specific function, care must be taken to make the chemical as non-toxic to humans and the environment as possible.
- **Design of safe auxiliaries and solvents:** The use of auxiliaries in processes must be avoided to the largest possible extent. Even in the circumstances where they need to be employed, they must be optimized to be as non-hazardous as possible.
- **Energy efficiency:** The amount of energy consumed by the process must be minimized to the maximum possible extent.
- **Incorporation of renewable feedstock:** The use of renewable feedstock and renewable raw materials must be preferred over the use of non-renewable ones.
- **Reduction in the generation of derivatives:** The unnecessary use of derivatives must be minimized since they tend to require

the use of additional reagents and chemicals, resulting in the generation of excess waste.

- **Incorporation of Catalysis:** To reduce the energy requirements of the chemical reactions in the process, the use of chemical catalysts and catalytic reagents must be advocated.
- **Designing the chemicals for degradation:** When designing a chemical product to serve a specific function, care must be taken during the design process to make sure that the chemical is not an environmental pollutant. This can be done by making sure that the chemical breaks down into non-toxic substances.
- **Incorporating real-time analysis:** Processes and analytical methodologies must be developed to the point that they can offer real-time data for their monitoring. This can enable the involved parties to stop or control the process before toxic/ dangerous substances are formed.
- **Incorporation of safe chemistry for the prevention of accidents:** While designing chemical processes, it is important to make sure that the substances that are used in the processes are safe to use. This can help prevent certain workplace accidents, such as explosions and fires. Furthermore, this can help develop a safer environment for the process to take place in.

2. EXAMPLES OF THE IMPACT OF GREEN CHEMISTRY

2.1. Use of Green Solvents

Solvents are mostly used in the liquid state, but they can exist in the solid and gaseous phases as well. solvents include water, ethanol, methanol, acetone, etc. Many chemical synthesis reactions that are carried out on an industrial scale require large amounts of chemical solvents. Furthermore, these solvents are also used industrially for degreasing and cleaning purposes. However, many traditional solvents that have been used for such purposes are toxic to human beings. Some such solvents are also known to be chlorinated. Green solvents, on the other hand, are generally less harmful to health and the environment and are preferably more sustainable. Ideally, solvents would be derived from renewable resources and biodegrade to an innocuous, often a naturally occurring product (Prat et al., 2013 and Sherman, 1998).

The advancement of green chemistry has brought many alternatives to these toxic solvents. The green solvents that are coming up as alternatives are known to be derived from renewable sources and are also known to be biodegradable. Thus, green chemistry has great potential to lower the toxicity of certain industrial environments by developing safer alternatives.

2.2. Development of Specialized Synthetic Techniques

The development of specialized synthetic techniques is an important aspect of green chemistry because it helps make chemical processes safer and more environmentally friendly. These techniques aim to reduce the use of toxic substances, minimize waste and improve the efficiency of chemical reactions. One of the most significant advancements in this field is the development of the olefin metathesis reaction, a powerful method used in organic chemistry for forming carbon-carbon double bonds. This reaction was developed by Robert H. Grubbs, Richard R. Schrock, and Yves Chauvin, who received the Nobel Prize in Chemistry 2005 for their contributions.

According to a review by Ryoji Noyori (2005), three major developments in green organic synthesis include the use of supercritical carbon dioxide as a green solvent, the use of aqueous hydrogen peroxide (H_2O_2) for cleaner oxidation reactions, and the use of hydrogen in asymmetric (enantioselective) synthesis. These approaches reduce the need for toxic solvents and produce fewer harmful by-products. Other applications of green chemistry include supercritical water oxidation (SCWO), dry media or solvent-free reactions, and on-water reactions, which help minimize environmental pollution.

Bioengineering also plays an important role in green chemistry by enabling the production of valuable chemicals through engineered microorganisms. For example, shikimic acid, a precursor used in the production of the antiviral drug Oseltamivir, is produced through bacterial fermentation by Roche. Another important approach is click chemistry, which involves simple and efficient reactions that produce minimal waste. The concept of green pharmacy, discussed by M. Baron (2012), applies green chemistry principles to pharmaceutical design and manufacturing.

2.3. Production of Hydrazine

Hydrazine was traditionally produced using the Olin Raschig Process, where ammonia reacts with sodium hypochlorite, producing sodium chloride as waste. Green chemistry introduced the Peroxide Process for Hydrazine Production, in which ammonia reacts with hydrogen peroxide. This method produces only water as a by-product and requires no additional solvents, making it a cleaner and more sustainable process (Schirmann, 2002).

$$NaOCl + 2\ NH_3 \rightarrow H_2N\text{-}NH_2 + NaCl + H_2O$$

In the greener peroxide process, hydrogen peroxide is employed as the oxidant and the side product is water. The net conversion follows:

$$2\ NH_3 + H_2O_2 \rightarrow H_2N\text{-}NH_2 + 2\ H_2O$$

Addressing principle #4, this process does not require auxiliary extracting solvents. Methyl ethyl ketone is used as a carrier for the hydrazine; the intermediate ketazine phase separates from the reaction mixture, facilitating workup without the need of an extracting solvent.

3. BIOREMEDIATION

"When microbes are utilized for cleaning a contaminated region, this process is called bioremediation. It is cost-effective and completely natural."

Bioremediation is a technique that uses microorganisms to clean polluted environments. Microorganisms such as bacteria, fungi and plants break down and consume harmful pollutants present in contaminated soil or water. These organisms convert toxic substances into less harmful products. As a result, bioremediation helps detoxify hazardous waste and restore polluted areas in an environmentally friendly way.

3.1. Principle of Bioremediation

The principle of bioremediation is based on harnessing the natural capabilities of certain microorganisms to break down and degrade environmental contaminants. The process involves the following key elements:

- **Contaminant Utilization:** Bioremediation uses specific microorganisms such as bacteria, fungi, and plants that can utilize pollutants as a source of food and energy. These microbes break down contaminants like oil, pesticides and solvents into simpler substances.
- **Microbial Conversion:** The microorganisms metabolize harmful chemicals and convert them into less toxic or harmless by-products such as water, carbon dioxide and small organic compounds, thereby reducing pollution.
- **Optimal Conditions:** Effective bioremediation requires suitable environmental conditions such as proper temperature, sufficient oxygen supply and necessary nutrients to support the growth and activity of microorganisms.
- **Environmental Amendments:** When natural conditions are not suitable, additional materials like molasses, vegetable oil, or aeration may be added to enhance microbial growth and improve the efficiency of the bioremediation process.
- **Timeframe:** The duration of bioremediation varies depending on factors such as the size of the contaminated area, concentration of pollutants, soil characteristics, temperature and whether the treatment is done in situ (at the site) or ex situ (outside the site). The process may take from several months to a few years.

3.2. Microrganisms used in bioremediation

Microorganisms play a vital role in nutrient cycles and maintaining biological balance. Bioremediation is the process of removing pollutants using microorganisms such as bacteria, fungi, algae, and yeast. These microbes can survive in extreme temperatures and use carbon for metabolism. Common bioremediating microbes include Achromobacter, Arthrobacter, Bacillus, Pseudomonas, Flavobacterium, Mycobacterium, Nitrosomonas and Xanthobacter.

3.3. Classification of Bioremediation

Bioremediation methods can be classified into different types based on the approach used and the location of the remediation process:

- **Natural Attenuation (Intrinsic Bioremediation):** This method relies on the natural ability of microorganisms already present in the environment to degrade contaminants. These microbes slowly break down pollutants in soil or water without human intervention, converting them into less harmful substances over time.
- **Biostimulation:** Biostimulation enhances the activity of existing microorganisms by adding nutrients such as fertilizers or other substances to the contaminated site. These nutrients stimulate microbial growth and increase the efficiency of pollutant degradation.
- **In Situ Bioremediation:** In situ bioremediation treats contaminated soil, groundwater, or sediments directly at the polluted site. This method avoids excavation and transportation of contaminated materials. Microorganisms degrade pollutants within the natural environment, making the process cost-effective and less disruptive.
- **Ex Situ Bioremediation:** Ex situ bioremediation involves removing contaminated materials from the original site and treating them elsewhere, such as in bioreactors or treatment facilities. This method is suitable for highly contaminated areas or locations where in situ treatment is not practical.

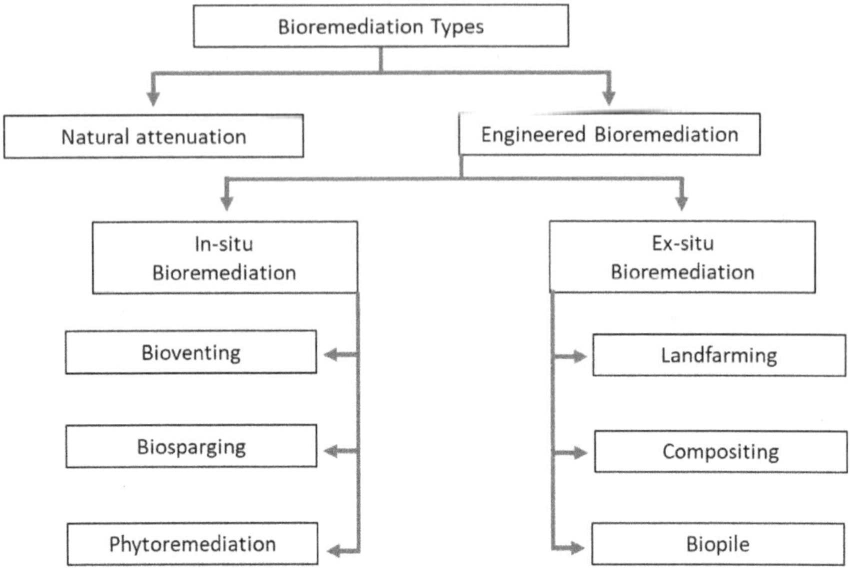

Figure. 1 Bioremediation types

3.4. Different Techniques of Bioremediation:

Phytoremediation is the process in which certain plants absorb pollutants from soil or water and convert them into less harmful substances. Plants such as Indian mustard, willow, poplar, and sunflower are commonly used. Another method, bioaugmentation, involves adding laboratory-cultured microbes to oxygen-rich contaminated sites to enhance the breakdown of pollutants.

3.5. Advantages of Bioremediation:

Bioremediation is a cost-effective and eco-friendly method for removing environmental pollutants using biological systems such as bacteria, microalgae, fungi and plants. Unlike conventional physicochemical treatments, it requires minimal infrastructure and is sustainable and scalable (Yuvraj, 2022). It helps control nutrient levels in water and prevents eutrophication. Organic pollutants can be biodegraded into carbon dioxide and water through oxidation processes (Duran et al., 2022; Singh et al., 2022). However, heavy metals cannot be degraded and are only transformed through oxidation or reduction, allowing partial removal (Kapahi et al., 2019). A major limitation of bioremediation is that the process is relatively slow (Hasanuzzaman et al., 2020).

4. WASTE MANAGEMENT

Waste refers to unwanted or unusable materials that are discarded after their primary use. It includes substances that are considered worthless, defective, or no longer useful. In contrast, a by-product is a secondary product of a manufacturing process that still has some economic value. In some cases, waste materials may become useful resources if new technologies or innovations increase their value. Examples of waste include municipal solid waste (household garbage), hazardous waste, wastewater such as sewage containing human excreta and surface runoff, and radioactive waste.

Effective waste management involves applying the 7R principle, which includes Refuse, Reduce, Reuse, Repair, Repurpose, Recycle, and Recover. The first two principles-refuse and reduce, focus on preventing waste generation by avoiding unnecessary purchases and minimizing consumption. Reuse and repair extend the life of existing products by using them multiple times or fixing damaged components. Repurpose and recycle ensure that materials are used again in new forms, reducing the need for raw resources. Recovery is considered the least efficient method and involves extracting energy from waste materials, such as burning waste to generate heat or electricity. Disposal of non-biodegradable waste in landfills is not considered a preferred waste management practice (Sankar & Ajith, 2015).

Waste management includes the collection, transportation, treatment, and final disposal of waste materials, along with monitoring, regulation, and the implementation of waste-related laws and technologies ("Environment Statistics", 2017). Waste can exist in solid, liquid, or gaseous forms, and each type requires different management methods. Waste management deals with various types of waste, including industrial, chemical, municipal, organic, biomedical and radioactive waste.

Improper waste handling can pose serious risks to human health and the environment (Giusti, 2009). Therefore, the main objective of waste management is to reduce these harmful impacts. A significant portion of waste management focuses on municipal solid waste generated from households, industries, and commercial activities (Wastes, 2023). Proper waste management is essential for developing sustainable and livable cities. However, it remains a major challenge in many developing countries because waste management systems can be expensive, often accounting for 20–50% of municipal budgets ("Solid Waste Management", 2020).

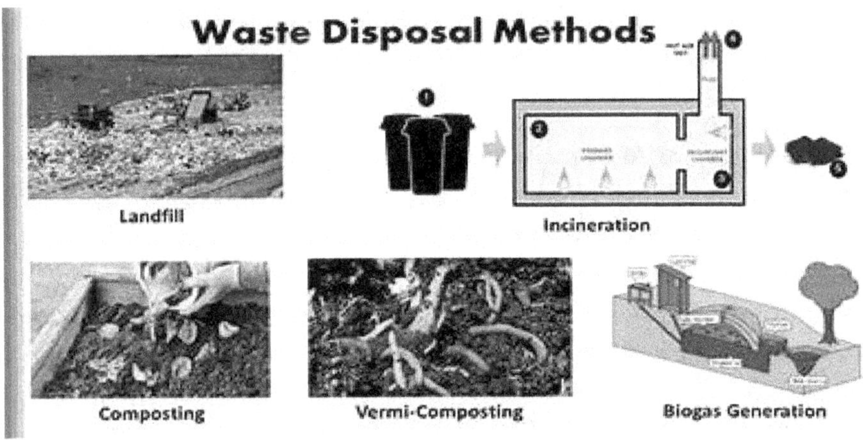

Figure. 2 Waste Disposal Methods

4.1. Practices of Waste Disposal Methods

Effective waste disposal is essential for protecting human health and the environment. Various methods are used to manage different types of waste in a safe and sustainable manner.

I. **Landfilling:** Landfilling is one of the most common waste disposal methods. In this method, waste is deposited in specially designed landfill sites and covered with layers of soil or other materials. The waste is compacted to reduce its volume and limit environmental impact. Modern landfills are engineered with protective liners and systems that collect and treat leachate, a liquid that forms when water passes through waste. They also capture methane gas produced during decomposition, which can be used for energy generation. Landfills are relatively cost-effective and can handle large amounts of waste.

II. **Composting:** Composting is a natural process in which microorganisms decompose organic waste such as food scraps, agricultural residues, and yard waste into a nutrient-rich material called compost. This compost can be used as a soil conditioner in agriculture and gardening. Composting reduces the amount of organic waste sent to landfills and helps improve soil fertility while decreasing dependence on chemical fertilizers. Composting can be carried out at household, community, or industrial scales.

III. **Incineration:** Incineration is a waste treatment process that involves burning solid organic waste at high temperatures. During this process, waste is converted into ash, heat, gases, and steam. Incineration significantly reduces the volume of solid waste by about 80–95% ("01-DMG", 2018). This method is useful for managing municipal solid waste and residues from wastewater treatment. Modern incinerators are equipped with pollution control systems that limit harmful emissions. Additionally, the heat produced during incineration can be used to generate electricity.

IV. **Recycling:** Recycling is a resource recovery method that involves collecting, processing, and reusing waste materials such as paper, glass, plastic, and metal. These materials are converted into raw materials for the production of new products. Recycling helps conserve natural resources, reduce energy consumption, and decrease the amount of waste sent to landfills. It also contributes to economic growth by creating new industries and job opportunities ("What is Recycling", 2020).

V. **Reuse:** Reuse refers to the practice of using an item or material again for the same or a different purpose without significant processing. Examples include reusing containers, repairing appliances, or donating used clothes. Reuse extends the lifespan of products, conserves resources, and reduces the demand for new materials and energy.

VI. **Sewage Treatment:** Sewage treatment is the process of treating wastewater generated from households, industries, and commercial activities. Treatment usually involves physical, chemical, and biological processes that remove contaminants and harmful microorganisms. Proper sewage treatment helps protect water bodies from pollution and prevents the spread of waterborne diseases. The treated water can often be safely discharged into rivers or reused for irrigation and other purposes.

VII. **Industrial Wastewater Treatment:** Industrial wastewater treatment involves processes designed to remove pollutants from wastewater generated by industries. Many industrial facilities, such as petroleum refineries, chemical plants, and petrochemical industries, have specialized treatment systems to ensure that wastewater meets environmental regulations before it is discharged into natural water bodies or sewer systems (Chobanoglous et al., 2003).

VIII. **Sewage Sludge Treatment:** Sewage sludge treatment deals with the management and disposal of sludge produced during sewage treatment. The main objective is to reduce its weight and volume to lower transportation and disposal costs while minimizing health risks. This is achieved through processes such as dewatering, thermophilic digestion, composting, or incineration. The selection of a treatment method depends on the amount of sludge generated and the cost-effectiveness of available disposal options.

4.2. Role of Responsible Waste Management in Sustainability

Responsible waste management is central to achieving sustainability goals for several reasons:

- **Resource Conservation:** Proper disposal methods like recycling and composting enable the recovery of valuable materials from waste, reducing the need for new resource extraction. This conserves natural resources and energy.
- **Reduction of Environmental Footprint:** Responsible waste management minimizes the environmental footprint of waste disposal operations, reducing their adverse impact on ecosystems and communities.
- **Promotion of Circular Economy:** By recycling and reusing materials, responsible waste management aligns with the principles of a circular economy, where resources are used efficiently and waste is minimized.
- **Enhanced Public Health:** It prevents the proliferation of disease vectors, the contamination of food and water sources, and exposure to harmful chemicals, contributing to improved public health.

To avoid the spread of public diseases and work towards a sustainable environment, companies have designed advanced and dependable products like bio-septic tanks and portable toilets. This is a way for the company to promote its motto that clean communities lead to healthier citizens.

REFERENCES

1. "01-DMG" (PDF). web.mit.edu. Archived (PDF) from the original on 19 June 2018.
2. "Environment Statistics" United Nations Statistics Division Archived from the original on 17 March 2017.Retrieved 3 March 2017.
3. "What is recycling". What is Recycling? 28 September (2020) – via Conserve Energy Future.
4. "Wastes". U.S. Environmental Protection Agency. 2 November 2017. Retrieved 19 August (2023.)

5. "Solid Waste Management". World Bank. Archived from the original on 30 September (2020).

6. Baron, M. (2012). "Towards a Greener Pharmacy by More Eco Design" (PDF). Waste and Biomass Valorization. **3** (4): 395–407.

7. Chobanoglous et al., (2003). Metcalf & Eddy Wastewater Engineering: Treatment and Reuse (4th ed.). McGraw-Hill Book Company. ISBN 0-07-041878-0.

8. Clark et al., (2012). "Green Chemistry, Biofuels, and Biorefinery". Annual Review of Chemical and Biomolecular Engineering. **3**: 183–207.

9. Duran N & Esposito E (2022). "Potential Applications of Oxidative Enzymes and Phenoloxidase-like Compounds in Wastewater and Soil Treatment: A Review". Applied Catalysis B: Environmental. **1** (2): 305–318.

10. Giusti, L. (1 August 2009)."A review of waste management practices and their impact on human health". Waste Management 29(8):2227-2239.

11. Introduction to In Situ Bioremediation of Groundwater (PDF). US Environmental Protection Agency. (2013). p. 30.

12. Jean-Pierre Schirmann & Paul Bourdauducq "Hydrazine" in Ullmann's Encyclopedia of Industrial Chemistry, Wiley Weinheim,2002.

13. Kapahi M. & Sachdeva S. (2019). "Bioremediation Options for Heavy Metal Pollution". Journal of Health and Pollution. **9** (24): 191203.

14. Mirza Hasanuzzaman, Majeti Narasimha Vara Prasad (2020). Handbook of Bioremediation. Academic Press. doi:10.1016/C2018-0-05109-9. ISBN 978-0-12-819382-2. S2CID 127409446.

15. Noyori, R. (2005). "Pursuing practical elegance in chemical synthesis". Chemical Communications (14): 1807–11.

16. Prat et al. (2013). "Sanofi's Solvent Selection Guide: A Step Toward More Sustainable Processes". Org. Process Res. Dev. **17** (12): 1517–1525.

17. Sankar, Ajith (2015). Environmental Management. New Delhi: Oxford University Press.

18. Sheldon, et al. (2007). Green Chemistry and Catalysis (PDF).

19. Sherman et al. (1998). "Solvent Replacement for Green Processing". Environ. Health Perspect. 106 (Suppl1): 253271.

20. Singh et al. (2019). "Role of Fungal Enzymes for Bioremediation of Hazardous Chemicals". Recent Advancement in White Biotechnology Through Fungi. Fungal Biology. Vol. 3. Cham: Springer International Publishing. pp. 237–256.

21. Yuvraj (2022). "Microalgal Bioremediation: A Clean and Sustainable Approach for Controlling Environmental Pollution". Innovations in Environmental Biotechnology. Vol. 1. Singapore: Springer Singapore. pp. 305–318.

Chapter - 14

CONVERSION OF CELLULOLYTIC WASTES INTO VALUABLE FERTILIZER USING "*EISENIA FETIDA*"

Sushobhit Varshney[1] and Priyanshi Saraswat[2]

[1,2] P G student, Department of Zoology, Sri Varshney College, Aligarh, 202001, Uttar Pradesh

Corresponding email: sushobhit552002@gmail.com

ABSTRACT

Vermicomposting is an eco-friendly and sustainable biotechnology that converts biodegradable waste into nutrient-rich organic manure through the activity of epigenic earthworms. The present study evaluates the efficiency of *Eisenia fetida* in converting kitchen waste, plant waste, and various cow dung-based mixtures into high-quality vermicompost. Five substrate combinations were prepared using cow dung alone and in combination with wood dust, bagasse, plant leaves, and flower waste. Additionally, selected kitchen and plant wastes were assessed as test substrates. The experiment was conducted over six months, and nutrient analysis was performed at 30, 60, and 90 days. Results revealed enhanced cocoon formation in kitchen waste substrates due to their high moisture and nutrient content, whereas plant waste substrates produced higher compost yield because of their fibrous carbon-rich composition. The vermicast obtained showed increased nitrogen, phosphorus, potassium, lignin, cellulose, humus content and a near-neutral pH (6.5–7.5). The findings highlight the significant role of *Eisenia fetida* in sustainable organic waste management and demonstrate the feasibility of transforming biodegradable residues into valuable biofertilizer for improving soil fertility and structure.

Keywords: *Vermicomposting, Eisenia fetida, kitchen waste, plant waste, cow dung mixtures, biomass conversion, organic manure, soil fertility.*

1. INTRODUCTION

Rapid urbanization, industrialization, and population growth have led to a significant increase in the generation of organic waste worldwide. Improper disposal of biodegradable waste contributes to environmental pollution, foul odor, groundwater contamination, and the emission of greenhouse gases such as methane. Therefore, sustainable and eco-friendly waste management strategies are urgently needed. Vermicomposting has emerged as an effective biological approach for converting organic waste into nutrient-rich manure using earthworms.

Figure 1. Eisenia fetida

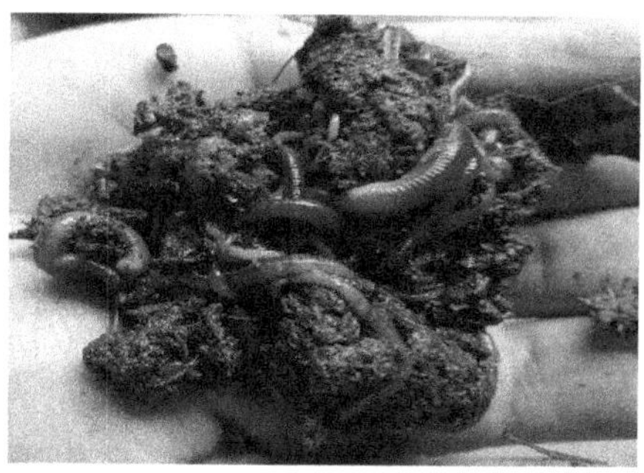

Figure 2. Vermicompost

Earthworms are often referred to as "ecosystem engineers" because of their vital role in soil formation, nutrient cycling, and organic matter decomposition. They belong to:

Phylum: Annelida

Class: Oligochaeta

Genus: *Eisenia*

Species: *fetida*

Among the various species used in vermicomposting, *Eisenia fetida* (commonly known as the red wiggler) is the most widely preferred. It is an epigenic, surface-dwelling earthworm characterized by rapid growth, high reproductive capacity, adaptability to organic waste environments, and tolerance to fluctuating environmental conditions. These biological traits make it highly efficient for large-scale composting systems.

Morphology and Biological Characteristics

The body of *E. fetida* is elongated, cylindrical, and segmented, typically measuring 1–18 cm in length. Each segment bears minute bristle-like structures called setae, which assist in locomotion and anchorage. The anterior end consists of the mouth, while the posterior end bears the anus. The moist body surface, maintained by mucus secretion, facilitates cutaneous respiration.

Earthworms possess a highly specialized digestive system capable of accelerating organic matter breakdown. They secrete important digestive enzymes, including:

- **Amylase** - Breaks down starch
- **Protease** - Breaks down proteins
- **Lipase** - Breaks down fats
- **Cellulase** - Breaks down cellulose

Additionally, the gut of *E. fetida* harbors a rich community of microorganisms such as bacteria, fungi, and actinomycetes, which synergistically enhance the decomposition process. Through mechanical grinding in the gizzard and enzymatic digestion in the intestine, organic residues are transformed into finely divided, stabilized casts known as vermicast.

Concept and Environmental Significance of Vermicomposting

Vermicomposting is a biological process in which earthworms convert biodegradable organic waste into stabilized, humus-like compost. During digestion, organic substrates are fragmented, chemically altered and enriched with microbial biomass, resulting in vermicompost, a nutrient-rich, eco-friendly biofertilizer. This product enhances soil structure, increases water-holding capacity, improves aeration, stimulates microbial activity and supplies essential nutrients such as nitrogen, phosphorus and potassium.

From an environmental standpoint, vermicomposting provides an efficient alternative to conventional waste disposal methods. By diverting organic waste from landfills, it significantly reduces methane emissions produced during anaerobic decomposition. Moreover, it minimizes reliance on chemical fertilizers, thereby lowering soil degradation and environmental toxicity. As a cost-effective and sustainable technology, vermicomposting supports organic farming systems and contributes to circular waste management practices.

In the present study, emphasis is placed on understanding the biological principles of vermicomposting, the operational mechanisms involved in waste conversion, and the agronomic as well as environmental benefits associated with the process. The objective is to evaluate the role of *E. fetida* in transforming kitchen and plant wastes into valuable organic manure, thereby promoting sustainable and environmentally friendly agricultural practices.

2. MATERIALS AND METHODS

2.1. Study Area and Duration

The experiment was conducted in the Zoology Laboratory of Shri Varshney College, Aligarh, Uttar Pradesh, India. The study was carried out over a period of approximately four months, from 27 January 2025 to 2 June 2025. The nutrient composition of vermicompost prepared from kitchen waste and plant waste was analyzed at three intervals: 30, 60 and 90 days.

2.2. Procurement of Earthworms

The epigenic earthworm species *Eisenia fetida* was procured from S.R.H. Organic Farm, Aligarh. A uniform quantity of 350 g of earthworms was introduced into each experimental grow bag.

2.3 Organic Substrates

Figure 3. Kitchen Waste

Figure 4. Plant leaves

Two categories of organic waste materials were used as substrates:

A. *Kitchen Waste*

- *Solanum tuberosum* (potato peels)
- *Raphanus sativus* (radish waste)
- *Brassica oleracea* (cabbage leaves)
- *Daucus carota* (carrot peels)
- *Camellia sinensis* (used tea leaves)

B. *Plant Leaves*

- *Ficus religiosa*
- *Azadirachta indica*
- *araca nucifera*
- *Ficus glomerata*
- *Delonix regia*

Fresh cow dung (6 kg) was used as a stabilizing agent and microbial inoculum.

2.4. Materials Used

- Two grow bags (12 × 12 × 24 inches)
- Jute bags (24 × 36 inches, 4 pieces)
- Gloves (3 pairs)
- Hand cultivator
- Water for moisture maintenance

2.5. Preparation of Substrates

The collected kitchen and plant wastes were initially moist; therefore, they were sun-dried for 10 days to reduce excess moisture and toxicity, making them suitable for earthworm inoculation.

Each grow bag was filled with approximately 30% organic waste (kitchen or plant waste) and 20% cow dung. The materials were mixed thoroughly to ensure uniform distribution. Moisture content was maintained at an optimal level by sprinkling water periodically.

The grow bags were covered with jute bags to maintain humidity and protect the worms from direct light. The setup was kept indoors under controlled laboratory conditions throughout the cultivation period.

2.6. Monitoring and Nutrient Analysis

The composting process was monitored regularly for earthworm activity and substrate decomposition. Samples of vermicompost were collected at predetermined intervals for nutrient analysis.

The nutrient composition of the compost was assessed based on the estimation of major components such as carbon, nitrogen, phosphorus, lignin, cellulose and humus content using standard analytical methods.

Calculation:

Component	Formula
Carbon	$C=\frac{100-\%ash}{1.8}$
Nitrogen	N= (1.4 *V*N)/W
Lignin	$L=\frac{L-A*100}{W}$
Cellulose	$C=\frac{Y-L*100}{W}$

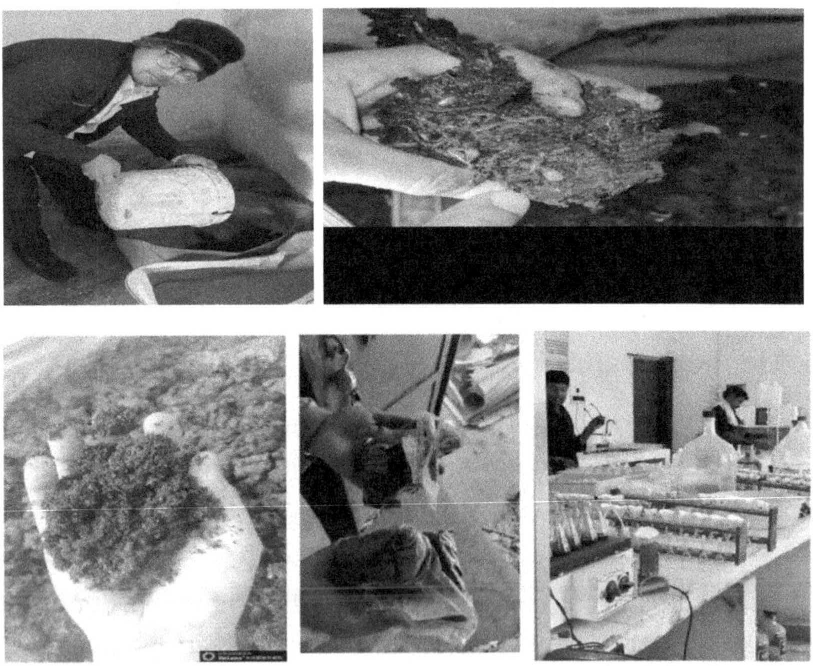

FINAL PRODUCT SAMPLE FOR TESTING LAB TESTING

Figure 5. Sequential process illustrating cow dung collection, earthworm transfer, production of vermicompost, and submission of the final product for laboratory analysis.

3. RESULTS

The values were recorded for two seasonal phases (January–March and April–June). The results are presented below in a systematic manner.

General Observations

Throughout the composting period, a gradual reduction in carbon, lignin, and cellulose content was observed, indicating active decomposition of organic matter. In contrast, nitrogen, phosphorus, and humus content increased over time, reflecting compost stabilization and nutrient enrichment. Carbon-rich materials such as sawdust and wood shavings acted as "brown" components, improving aeration and structural stability of the composting medium and preventing compaction.

Nutrient amount (g/kg) in Kitchen Waste

Component	Month	30 day	60 day	90 day
Carbon	Jan- Mar	33.28±0.62	22.17±0.13	16.41±0.18
	Apr-Jun	33.62±0.54	22.12±0.17	16.31±0.28
Nitrogen	Jan- Mar	1.28±0.023	1.78±0.021	2.06±0.023
	Apr-Jun	1.31±0.017	1.79±0.011	2.18±0.016
Phosphorus	Jan- Mar	79.38±12.14	85.56±5.41	88.39±11.61
	Apr-Jun	75.86±4.41	82.59±5.91	86.28±11.61
Lignin	Jan- Mar	4.17±0.45	3.86±0.49	3.83±0.52
	Apr-Jun	3.94±0.68	3.56±0.38	3.49±0.52
Cellulose	Jan- Mar	31.32±0.76	28.59±0.45	27.62±1.35
	Apr-Jun	30.19±0.81	29.05±0.74	27.15±1.28
Humus	Jan- Mar	3.74±0.42	4.32±0.43	4.92±0.22
	Apr-Jun	6.82±0.11	7.46±0.29	8.16±0.06

Nutrient amount (g/kg) in plant waste

Compound	Month	30 days	60 day	90 day
Carbon	Jan-Mar	13.99±0.28	10.80±0.83	8.81±0.54
	Apr-Jun	13.48±0.63	10.72±0.49	8.72±0.43
Nitrogen	Jan-Mar	3.21±0.048	3.72±0.044	4.13±0.58
	Apr-Jun	3.29±0.057	3.76±0.023	4.21±0.79
Phosphorus	Jan-Mar	71.76±10.29	73.32±2.80	76.42±18.9
	Apr-Jun	64.48±9.36	66.34±6.20	69.89±5.41
Lignin	Jan-Mar	6.52±2.14	6.17±0.67	5.59±0.33
	Apr-Jun	6.09±0.78	5.91±1.26	5.68±1.43
Cellulose	Jan-Mar	32.62±0.82	30.12±0.34	29.46±0.59
	Apr-Jun	30.92±0.67	27.98±1.07	24.99±2.63
Humus	Jan-Mar	3.46±0.39	3.69±0.51	4.35±0.59
	Apr-Jun	5.56±0.42	6.32±0.28	7.54±0.38

Comparative Analysis

- Both kitchen and plant waste showed a consistent reduction in carbon, lignin and cellulose over time, indicating effective decomposition.
- Nitrogen and phosphorus content increased progressively, reflecting nutrient enrichment.
- Humus content increased in both substrates, confirming compost stabilization.
- Plant waste exhibited relatively higher nitrogen values compared to kitchen waste, while kitchen waste showed substantial phosphorus enrichment.

Overall, the results demonstrate that *Eisenia fetida* effectively converted both kitchen and plant wastes into nutrient-rich vermicompost within 90 days.

4. DISCUSSION

The present study evaluated the changes in nutrient composition of kitchen waste and plant waste during vermicomposting at 30, 60, and 90 days, and correlated these changes with earthworm growth, cocoon formation, and compost production.

Changes in Nutrient Composition During Vermicomposting

A consistent decline in carbon content was observed in both substrates over time. In kitchen waste, carbon decreased markedly from the 30th to the 90th day in both seasonal batches (Jan–Mar and Apr–Jun). A similar declining trend was recorded in plant waste. The reduction in carbon content indicates active microbial degradation and mineralization of organic matter, as carbon is released in the form of CO_2 during respiration. This decline confirms effective decomposition during the vermicomposting process.

In contrast, nitrogen content showed a steady increase in both substrates over time. Kitchen waste exhibited a gradual rise in nitrogen concentration from 30 to 90 days, while plant waste showed comparatively higher nitrogen values overall, with a clear incremental pattern across all time intervals. The increase in nitrogen may be attributed to the concentration effect caused by carbon loss, along with the addition of nitrogenous excretory products, mucus, and microbial biomass associated with earthworm activity.

Phosphorus content also showed an increasing trend with time in both wastes. The enhancement of phosphorus availability is likely due to mineralization and microbial solubilization processes occurring in the earthworm gut and surrounding substrate. Earthworm activity enhances phosphatase enzyme production, which converts insoluble phosphorus into forms more available in the final compost.

Lignin and cellulose contents gradually decreased over the composting period in both substrates, indicating breakdown of complex structural compounds. The reduction was more pronounced in plant waste due to its initially higher fibrous composition. This decrease reflects the combined action of microorganisms and earthworms in degrading resistant organic fractions.

Humus content increased progressively in both substrates, particularly at later stages (90 days). The formation of humus indicates stabilization and maturation of compost. The higher humus values observed at 90 days confirm improved compost quality and greater nutrient stabilization.

Relationship with Cocoon Formation and Compost Production

The biological observations showed that the 60-day interval of kitchen waste resulted in the highest cocoon formation. This period likely represented an optimal balance between nutrient availability and substrate stability. At 60 days, sufficient decomposition may have occurred to make nutrients readily available without excessive stabilization, thus creating favorable conditions for earthworm reproduction.

In contrast, the 90-day interval of plant waste exhibited both maximum cocoon formation and highest compost production. By this stage, plant waste had undergone substantial degradation of lignin and cellulose, improved nutrient concentration (particularly nitrogen and phosphorus), and enhanced humus formation. These factors likely created a stable and nutrient-rich environment conducive not only to reproduction but also to efficient compost maturation.

Kitchen waste, being rich in easily degradable organic matter and moisture, appears to favor reproductive performance at intermediate stages (60 days). Plant waste, due to its fibrous and carbon-rich nature, supports sustained decomposition and higher compost yield at later stages (90 days).

Influence of Substrate Type and Duration

The findings demonstrate that both substrate type and composting duration significantly influence vermicomposting efficiency. Kitchen waste promotes rapid nutrient transformation and early reproductive response, whereas plant waste requires longer decomposition time but ultimately produces more stabilized and nutrient-rich compost.

Seasonal variations (Jan–Mar and Apr–Jun) showed minor differences in nutrient values, possibly due to environmental factors such as temperature and moisture. These parameters can influence microbial activity and earthworm metabolism, thereby affecting decomposition rate and reproductive performance.

CONCLUSION

Overall, the study confirms that vermicomposting significantly improves nutrient composition over time, as indicated by reduced carbon, decreased lignin and cellulose, and increased nitrogen, phosphorus, and humus content. The 60-day period was optimal for cocoon production in kitchen waste, while 90 days was most effective for plant waste in terms of both reproduction and compost yield.

These results highlight the importance of selecting appropriate substrate type and composting duration to maximize earthworm activity and compost quality in vermicomposting systems.

REFERENCES

1. Chekwube Enebe, M., & Erasmus, M. (2023). Mediators of biomass transformation – A focus on the enzyme composition of the vermicomposting process. *Environmental Challenges, 12*, 100732.

2. Kostecka, J., Garczyńska, M., Podolak, A., Pączka, G., & Kaniuczak, J. (2018). Kitchen organic waste as material for vermiculture and source of nutrients for plants. *Journal of Ecological Engineering, 19*(6).

3. Lankester, E. (1864). The anatomy of the earthworm. *Journal of Cell Science, 2*(16), 258–268.

4. Manaf, L. A., Jusoh, M. L. C., Yusoff, M. K., Ismail, T. H. T., Harun, R., Juahir, H., & Jusoff, K. (2009). Influences of bedding material in vermicomposting process. *International Journal of Biology, 1*(1), 81.

5. Neuhauser, E. F., Kaplan, D. L., Malecki, M. R., & Hartenstein, R. (1980). Materials supporting weight gain by the earthworm *Eisenia foetida* in waste conversion systems. *Agricultural Wastes, 2*(1), 43–60.

6. Ratnasari, A., Syafiuddin, A., Mehmood, M. A., & Boopathy, R. (2023). Review of the vermicomposting process of organic and inorganic waste in soils: Additives effects, bioconversion process, and recommendations. *Bioresource Technology Reports, 21*, 101332.

7. Siddique, J., Khan, A. A., Hussain, I., & Akhter, S. (2005). Growth and reproduction of earthworm (*Eisenia fetida*) in different organic media. *Pakistan Journal of Zoology, 37*(3), 211–215.

8. Singh, S., Sharma, A., Khajuria, K., Singh, J., & Vig, A. P. (2020). Soil properties changes earthworm diversity indices in different agro-ecosystem. *BMC Ecology, 20*(1), 27.

9. Wani, K. A., & Rao, R. J. (2013). Bioconversion of garden waste, kitchen waste and cow dung into value-added products using earthworm *Eisenia fetida*. *Saudi Journal of Biological Sciences, 20*(2), 149–154.

Chapter - 15

ECOTOXICOLOGICAL EFFECTS OF ENVIRONMENTAL POLLUTANTS ON BIOTA AND ECOSYSTEM DYNAMICS

Dr. Sonam Shakya[1], Dr. Barkha Shakya[2] and Dr. Ishaat M. Khan[1]

[1]Department of Chemistry, Aligarh Muslim University, Aligarh 202002, India

[2]Department of Zoology, Aligarh Muslim University, Aligarh 202002, India

***Corresponding email: sonamshakya08@gmail.com**

ABSTRACT

The high level of pollution, especially environmental pollution, has become one of the big threats to global biodiversity and the stability of ecosystems. This chapter reviews the toxic chronic effects of a variety of environmental pollutants (heavy metal atomics, pesticides, microplastics), industrial chemicals, and atmospheric pollutants, on organisms and ecosystems. It covers mechanisms of toxicity, bioaccumulation, biomagnification, and the cascade effects these substances exert on ecological processes. That includes sensitive species, trophic interactions, ecosystem services and issues related to environmental risk assessments.

Keywords: Environmental Pollutants; Ecotoxicology; Bioaccumulation and Biomagnification; Ecosystem Dynamics

1. INTRODUCTION

Over the past century, the world's agriculture, industry, and urban centres became increasingly productive at the same time as new technologies and practices were introduced. The use of synthetic chemicals and other pollutants has been steadily increasing and now these pollutants include pesticides, heavy metals, microplastics, among other things (Edwards, 2002; Belden, 2020). Persistent organic pollutants are all now accepted to be some of the world's greatest dangers to its biodiversity and balance of ecosystems. Most of the pollutants mentioned previously are bio-accumulative and persistent. This means they do not break down and are passed on to the most untouched and sensitive ecosystems on the planet (Kazlauskiene et al., 2012).

Ecotoxicology studies the effects that chemical and biological contaminants have on living organisms with respect to their population and community and therefore serves as a reminder of the elusive connection between ecology and toxicology (Pesce et al., 2023; Brooks, 2022). Every single organism can be negatively affected by pollutants (Karimi et al., 2020). The pollutant however does not have to be present in the entire ecosystem– it can be located within one organism and still create such damage because of other mechanisms known as bioaccumulation and biomagnification. Biomagnification occurs when predators at the top of the food chain amplify the concentration of toxins in their bodies and subsequently pass them on to their offspring. Other irreversible harm that can be caused include oxidative stress, endocrine disruption and epigenetic modification, which through many generations can lower survivability and fitness (Chormare et al., 2022; Ali et al., 2019; Obiakor et al., 2017)

For the purposes of global concentration policies, it is very crucial to understand the methods, effects, and mechanisms of pollution because it allows aiming at gaps to improve decision-making effectiveness. The purpose of this chapter is to analyze the oft unheard effects of ecotoxicology.

2. ENVIRONMENTAL POLLUTANTS

Natural habitats have become contaminated by the wide spread of synthetic chemicals and industrial activities (Suzuki et al., 2020; Verhaar et al., 1992). Pollutants are the generally harmful substances that we introduce into the environment which have negative impacts on living organisms and the biophysical environment (Shetty et al., 2023). These pollutants are often persistent in the environment, undergo long-range transport, and bioaccumulate in biological tissues, making them a chronic risk to wildlife and humans.

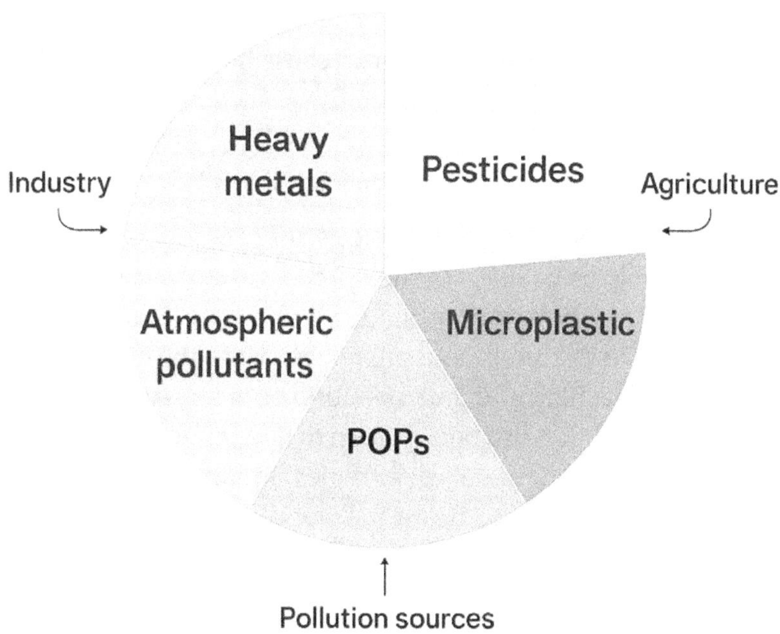

Figure 1. Major types of environmental pollutants.

Following are the different types and sources of environmental pollutants:

2.1. Heavy Metals

Among these, heavy metals (HMs) like mercury (Hg), cadmium (Cd), arsenic (As), and lead (Pb) are emitted from several environmental anthropogenic sources, e.g., in mining practices, metal smelting, coal firing, and industrial wastewater (Wasi et al., 2013). Heavy metals, unlike organic pollutants, will not degrade in the environment, and hence they may remain indefinitely. In aquatic systems, they can bind to organic and inorganic ligands to form organo–metal complexes that enhance their mobility and bioavailability. Heavy metals induce neurological damage, renal dysfunction, impaired reproduction, and immune suppression in organisms upon chronic exposure. In plants, heavy metals disrupt photosynthesis and nutrient uptake mechanisms, resulting in decreased growth and productivity (Förstner et al., 1973). One troubling example of this phenomenon is the biomagnification of mercury through aquatic food webs, which results in elevated concentrations in top predators, such as fish, birds, and mammals that can cause harm (Mohammad Ali et al., 2021; Cook, 1977).

2.2. Pesticides

Pesticides are artificially made chemicals that target pests but affect non-target organisms because of their broad-spectrum activity. Classes used often include organophosphates, pyrethroids, carbamates and neonicotinoids. These compounds are released into the environment by spraying, runoff, leaching and drift, polluting both terrestrial and aquatic habitats. Pesticides have also been shown to have effects on fish survival, behavior, and developmental deformities in aquatic environments (Edwards, n.d.). In addition to being the world's leading pollinator, bees and butterflies are extremely sensitive to pesticides; neonicotinoids have been shown to interfere with navigation, foraging and colony health. Even at low pesticide concentrations, lethal and sublethal effects in amphibians occur. Moreover, pests can become resistant to prolonged pesticide treatment, resulting in the use of increasingly nefarious chemicals that can cause greater ecological damage (Fisken, 1972).

2.3. Microplastics

Microplastics are recently disclosed global contaminants released via fragmentation of larger plastic materials, or they may be directly introduced through consumer products and industrial procedure.

These particles are ubiquitous among ecosystems, from the deep oceans and polar regions to freshwater rivers and agricultural soils. Different organisms from plankton to whales ingest microplastics, usually mistaking them as food (Xiang et al., 2022). Ingested, they create gut obstruction, impaired feeding efficiency and energetic stress. Microplastic also serve as vectors for hydrophobic contaminants, such as PCBs (Polychlorinated Biphenyls) and PAHs (Polycyclic Aromatic Hydrocarbons), that tend to adsorb on their surfaces (Figure 2). These contaminants can subsequently make their way into the food web and bioaccumulate. Other studies have shown that microplastics change the structure of microbial communities and influence soil health by disrupting water retention and root development (Alimba et al., 2019).

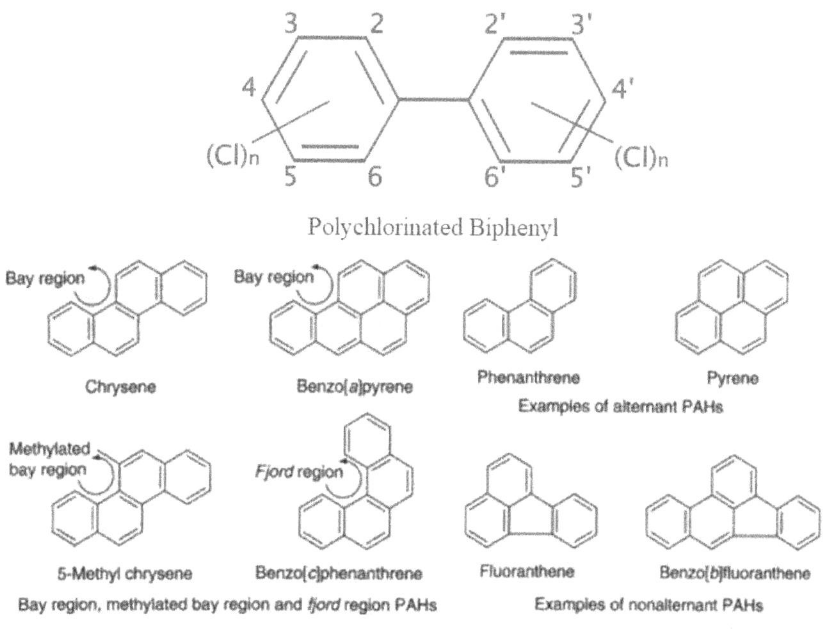

Figure 2. Molecular structures of polychlorinated biphenyls (PCBs) and polycyclic aromatic hydrocarbons (PAHs).

2.4. Industrial Chemicals

Persistent organic pollutants (POPs) are a group of industrial chemicals with long environmental half-lives, lipophilicity, and the potential to bioaccumulate in living organisms. Persistent organic pollutants (POPs)- including polychlorinated biphenyls (PCBs), dioxins, and polycyclic aromatic hydrocarbons (PAHs)- are emitted from waste incineration, fossil fuel combustion, and the production of pesticides and solvents. These chemicals disrupt endocrine signalling pathways and cause endocrine disruption. To the adverse effects include reproductive failures, immunotoxicity, developmental delays, and carcinogenesis. POPs travel around the globe in the air and water, contaminating distant ecosystems, including those in the Arctic. Their presence in human tissues, breast milk and wildlife is driving the need for tight international controls and safe disposal processes (Vandana et al., 2022).

2.5. Atmospheric Pollutants

Fossil fuel combustion ultimately results in atmospheric pollutants, but we also know that vehicular emissions, industrial processes, and biomass burning can emit atmospheric pollutants. Important pollutants are ozone (O3), nitrogen oxides (NOx), sulphur dioxide (SO2), carbon monoxide (CO), and particulate matter (PM). Ground-level ozone is a potent oxidant that injures plant tissues, inhibits photosynthesis, and diminishes crop yields. NOx and SO2 from fossil fuels contribute to acid rain, which lowers soil pH, leaches essential nutrients, and acidifies water bodies, damaging aquatic life. Metals, black carbon and organic toxins can present in particulate matter and airborne dust which can be inhaled by terrestrial and aquatic organisms potentially resulting in respiratory and cardiovascular effects. Atmospheric deposition also enables airborne contaminants to be deposited onto soil and water, extending their ecological impact. Climate change also compounds air pollution, increasing stress on species and ecosystems (Muñoz et al., 2015; Hogg et al., 2009).

3. MECHANISMS OF ECOTOXICITY

3.1. Bioaccumulation and Biomagnification

The progressive accumulation of pollutants to an organism from an environmental medium (such as water, soil, or air) through absorption is referred to as bioaccumulation. Lipophilic molecules, like PCBs and mercury compounds, tend to bioaccumulate because they are poorly soluble in water and bind strongly to fatty tissues. Biomagnification, in contrast, refers to the increasing concentration of pollutants as they are transferred through a food web from one trophic level to another. This was especially observable in aquatic ecosystem, where the highest levels of contaminants tend to accumulate in top predators, such as large fish, birds, and marine mammals. Apex species undergo consequential physiological and reproductive changes and the integrity of them as well as the larger network becomes compromised (Streit, 1992).

3.2. Oxidative Stress

Pollutants can have toxic effects on biological systems through various mechanisms, with one of the most important being the induction of oxidative stress. When heavy metals, PAHs, pesticides, and other pollutants enter the cells, they can induce overproduction of reactive oxygen species (ROS). ROS are highly reactive molecules that affect the cellular components like DNA, proteins, and lipids. Oxidative stress generally elicits a defence response that toll the expression of antioxidant systems like superoxide dismutase (SOD) and catalase (Valavanidis et al., 2006). But, excessive production of ROS exceeds the capacity of the cellular defence mechanisms leads to cellular damage, resulting apoptosis, inflammation or even disease states i.e. cancer, neurodegeneration. Chronic oxidative stress also disrupts physiological functions such as reproduction, immunity, and energy metabolism (He et al., 2023).

3.3. Endocrine Disruption

Human exposure to endocrine-disrupting chemicals (EDCs), agents that interfere with normal hormone action, is a growing concern.

Such substances include pesticides, plasticizers (eg, bisphenol A), pharmaceuticals and industrial by-products. EDCs can imitate natural hormones such as oestrogens or androgens, obstruct hormone receptors, or interrupt hormone production and metabolism. In wildlife, exposure to EDCs has been associated with feminization of male fish, altered reproductive behaviour in amphibians, and developmental defects in birds and mammals. These disruptions can have population-wide consequences, including unbalanced sex ratios and decreased fertility (Goksøyr, 2006). In addition, the so-called endocrine effects occur at very low pollutant concentrations, which raises concerns about safety threshold concentrations and adds more complexity to the regulatory process.

3.4. Genetic and Epigenetic Effects

They can cause molecular changes affecting genetic integrity and gene expression. PAHs, benzene, heavy metals and other genotoxic compounds can result in DNA damage, including strand breaks, mutations, and chromosomal aberrations. These genetic changes can lead to developmental defects, cancer, or inherited diseases. Beyond mutating genes, pollutants can lead to epigenetic modifications, inheritable traits in which the gene function changes while the sequence itself is not altered. These include DNA methylation, histone modification, and non-coding RNA expression changes. Epigenetic modifications can modulate phenotypic plasticity, stress responses, and adaptation, but also exert deleterious long-lasting effects on population health and evolutionary potential (Vandegehuchte et al., 2011).

4. IMPACT ON INDIVIDUAL ORGANISMS AND POPULATIONS

The short and long-term effects of environmental pollutants can be substantial at the organismal level and on the population as a whole. For the individual, exposure to toxicants can cause lethal effects, such as death, or sublethal effects, including impaired reproduction, altered behaviour, weakened immunity, or delayed growth (Bergek et al., 2012). These sublethal effects, while not immediately deadly, can strongly diminish an organism's fitness and increase its susceptibility to predation, disease or environmental stress (Figure 3).

Pollution is another powerful driver of population declines across many species. Population viability is at risk when reproductive success decreases and mortality rates increase for long periods of time. Long-lived species with low birth rates: large mammals or apex predators, for example: are particularly at risk because they do not rebound quickly from drops in population. Endocrine disruption or genotoxic stress can result in reproductive failures that not only affect individuals, but also the population (e.g., sex ratio skewing or loss of genetic variability) resulting in demographic imbalances that can destabilize population dynamics further (Vasseur et al., 2006).

Certain species are particularly susceptible to pollutants because of their physiological properties or ecological niches. Amphibians, for example, have porous skin and live part of their life cycle in water, which makes them very vulnerable to contaminants in drinking water. Louse flies, molluscs and aquatic insects also tend to be sensitive to changes in water chemistry, and are among the first to show signs of pollution stress. This makes them important bioindicators in environmental tracking programs as their loss will serve as an early warning of ecological disruption (Samuel et al., 2023).

Overall, the cumulative impact of pollutants on individual health and population dynamics can lead to cascading ecological consequences, including altered community composition and loss of ecosystem services. Understanding these effects is vital for designing effective conservation strategies and ensuring the long-term sustainability of affected ecosystems.

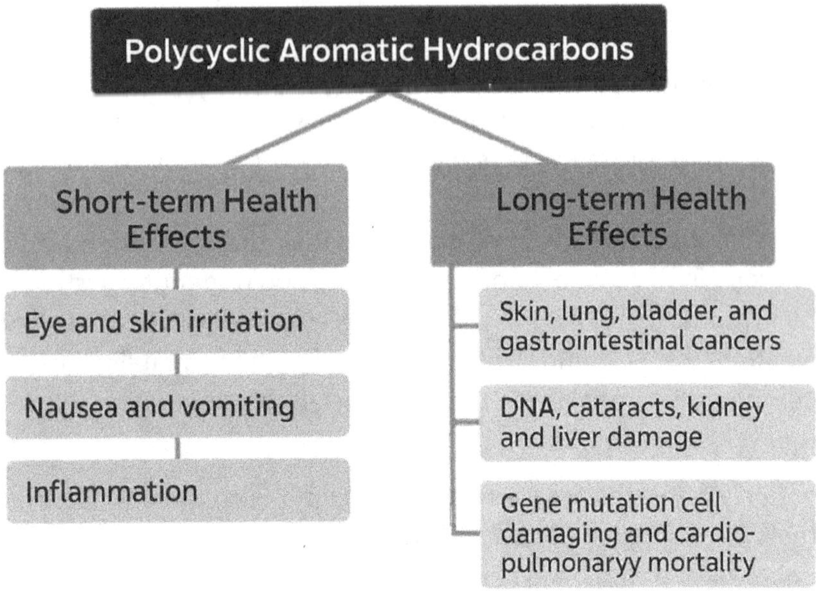

Figure 3. Short-term and long-term health effects of polycyclic aromatic hydrocarbons (PAHs).

5. EFFECTS ON COMMUNITY STRUCTURE AND ECOSYSTEM DYNAMICS

Environmental pollutants not only affect individual species, but also influence ecological community structure and functioning. Pollutants modify community dynamics primarily through the disruption of trophic interactions. When a predator species declines, for instance due to the bioaccumulation of toxicants, it's prey growth will not be limited by top-down controls and could run unchecked resulting in overgrazing or overexploitation of resources. By contrast, a reduction in primary producers through toxicity can topple entire webs, causing trophic cascades. If pollinators or seed-dispersing animals are adversely impacted by toxic exposures, mutualistic relationships, e.g., for pollination and seed dispersal, may fail (Menezes-Oliveira et al., 2014).

One consequence of stress induced by pollution is a decrease in biodiversity, in terms of species richness (the number of species) and evenness (the distribution of individuals across species). Initially, sensitive species are lost, followed by a few tolerant or opportunistic species that dominate the modified ecosystem. This loss of biodiversity reduces ecological resilience; the capacity of our ecosystems to absorb disturbances and responds and, making them more vulnerable to other stressors, from climate change to invasive species. Ecological communities may split apart, looser linkages can cause less specialization for a given niche and homogenization of habitat (Rutgers, 2008).

The disruption of ecosystem services and the natural processes that benefit all human societies is another major consequence of environmental pollution. For instance, the collapse of insect pollinators from the use of pesticides directly threatens agricultural productivity. Wetlands and aquatic systems are also affected by contaminants that interfere with their ability to filter water and cycle nutrients, thereby reducing water quality and impacting fisheries. Soil contamination can affect microbial communities and reduce nutrient availability to plants, and thus potentially affect crop yields and forest regeneration. This degradation of these services has economic and social consequences of the higher importance, especially for the communities that rely directly on natural resources (Clements et al., 2009)diversity, community composition, and species interactions. Recent discoveries that species diversity is positively associated with ecosystem stability, recovery, and services have made a community-level perspective on ecotoxicology more important than ever. Community ecotoxicology must explicitly consider both present and impending global change and shift from a purely descriptive to a more predictive science. Greater consideration of the ecological factors and threshold responses that determine community resistance and resilience should improve our ability to predict how and when communities will respond to, and recover from, xenobiotics. A better understanding of pollution-induced community tolerance, and of the costs of this tolerance, should facilitate identifying contaminant-impacted communities, thus forecasting the ecological consequences of contaminant exposure and determining the restoration effectiveness. Given the vast complexity of community ecotoxicology, simplifying assumptions, such as the possibility that the approximately 100,000 registered chemicals could be reduced to a more manageable number of contaminant classes with similar modes of action, must be identified and validated. In addition to providing a framework for predicting contaminant fate and effects, food-web ecology can help to identify communities that are sensitive to contaminants, contaminants that are particularly insidious to communities, and species that are crucial for transmitting adverse effects across trophic levels. Integration of basic ecological principles into the design and implementation of ecotoxicological research is essential for predicting contaminant effects within the context of rapidly changing, global environmental conditions (Clements and Rohr, 2009).

In sum, pollutants can fundamentally alter community composition, ecological interactions, and essential ecosystem functions. Addressing these impacts requires a systems-level understanding of ecological networks and a commitment to pollution prevention, habitat restoration, and the maintenance of ecological integrity.

6. CASE STUDIES

6.1. Mercury in Aquatic Food Webs

Mercury is a heavy metal that enters and poses a serious danger to aquatic ecosystems due to its extremely toxic properties. One of its most toxic forms, methylmercury, is shaped by microbiological activity within aquatic sediments, readily absorbed by organisms. It bioaccumulates in the tissue of aquatic organisms and biomagnifies in the food chain, resulting in a higher concentration in predator fish and piscivorous birds. Mercury can damage neurological development, reproductive function and behaviour at chronic exposure levels found in fish, all leading to population declines. In birds like eagles and ospreys, mercury poisoning leads to lower hatching success and deformities in chicks. These impacts not only jeopardize biodiversity but also affect human health via the ingestion of polluted fish (Pedro et al., 2017; Lavoie et al., 2013).

6.2. Neonicotinoids and Pollinator Decline

Neonicotinoids are systemic insecticides that have attracted worldwide concern given their detrimental effects on pollinators. Those compounds are absorbed by plants and spread throughout their tissues, including in pollen and nectar. Pollinators like honeybees and bumblebees that visit treated plants consume such chemicals, disrupting their nervous systems. These may result in sublethal effects such as impaired navigation, reduced foraging efficiency, weakened immune response and disrupted colony dynamics. Extensive neonicotinoid use has been shown to associate with large declines in pollinators observed in the field, and such declines threaten food security as bees are crucial for the pollination of many crops. The environmental exposure and ecological impact are also amplified due to their persistence in soil and water (Wood et al., 2017).

6.3. Microplastics in Marine Ecosystems

Microplastics are a ubiquitous and increasing threat to marine ecosystems. These particles, either resulting from the breakdown of larger plastic debris or discharging from direct sources, such as cosmetic products, pervade almost all marine habitats. Microplastics end up in marine life from plankton to molluscs to fish to even the sea whales that eat them, mistaking them for food. After ingestion, microplastics can puncture internal organs, obstruct digestive tracts, and induce starvation. They also act as carriers of chemical contaminants and pathogens, exacerbating toxic exposure and the risk of disease. Microplastics can also interfere with feeding and reproduction, disrupting growth rates and survival. The ubiquity of seafood makes human exposure and long-term ecological ramifications a major concern (Gola et al., 2021).

6.4. Oil Spills and Coastal Ecosystems

Oil spills are some of the most devastating types of environmental pollution, with coastal and marine ecosystems bearing the brunt. Complex mixtures of hydrocarbons within crude oil and refined petroleum products are toxic to marine life. Once spilled, oil floats atop the surface of water, inhibiting oxygen exchange and suffocating marine organisms. Oil's toxic constituents, particularly polycyclic aromatic hydrocarbons (or PAHs), can be absorbed across gills or ingested, resulting in tissue damage and immunosuppression, and reproductive failure of fish, seabirds, and marine mammals. Mangroves, salt marshes, and coral reefs are key contributors to coastal health, but are particularly sensitive to disturbance because of their narrow ability to recover from physical and chemical disturbance. Oil can linger in sediments for decades, extending exposure and preventing ecosystems from bouncing back. High-profile disasters such as the Deepwater Horizon spill in the Gulf of Mexico illustrate the far-reaching ecological and socioeconomic impacts of oil contamination (Zhang et al., 2019).

7. ENVIRONMENTAL RISK ASSESSMENT AND MANAGEMENT

Environmental risk assessment and management play a crucial role in understanding, mitigating, and regulating the impacts of pollutants on ecological and human health. Environmental monitoring, which is a systematic measure of pollutant concentrations in air, water, soil, and biota, is thus an essential part of risk assessment. It is useful for finding pollution hotspots, determining temporal trends and compliance with regulatory limits. Biomonitoring, which focuses on the biological responses of sentinel species to contaminants, complements this and provides early-warning indicators of ecosystem stress. The most often used organisms are mussels, lichens, and earthworms due to their sensitivity to contaminants and ecological relevance (Rathi et al., 2021).

Alongside monitoring initiatives, implementing regulatory frameworks is crucial to curtail pollution threats. International agreements such as the Stockholm Convention on Persistent Organic Pollutants and regional initiatives like the European Union's REACH regulation have established standards for the identification, restriction and elimination of toxic compounds. In addition, national environmental agencies reinforce these activities by providing permissible exposure limits, conducting environmental impact assessments (EIAs), and raising public awareness (Efroymson et al., 2001). But challenges persist through patchy enforcement, newly emerging contaminants where little data exists, and lack of monitoring infrastructure in many regions of the world.

When pollutants are detected, remediation strategies must be designed and executed to mitigate their concentration and harm. Deploying micro-organisms to either degrade or sequester the toxic chemicals is a low-cost and eco-friendly alternative, a process known as bioremediation. From soil and water bodies, pollutants are extracted, sequestered, or degraded using plants. Depending on the nature and persistence of contamination, more advanced technologies are employed, including activated carbon adsorption, chemical degradation, and nanomaterial-based treatment processes, to remove these contaminants effectively (Tang et al., 2019). Integrated management approaches that integrate remediation with the restoration of habitat and the prevention of pollution are being increasingly advocated for to achieve lasting ecosystem recovery.

Overall, environmental risk assessment and management involve the cooperation of various disciplines, the use of strong scientific evidence, and adaptable regulatory frameworks to safeguard ecosystems from the increasing pollution burden. With the expansion of industrial activities and the emergence of new pollutants, it is imperative to continue to innovate, act affirmatively, and remain vigilant in order to maintain ecological and human well-being.

8. CONCLUSION

Environmental pollution has multifaceted and far-reaching impacts at biota and ecosystem functioning levels. The ultimate goal of studying ecotoxicological mechanisms and ecological responses is to develop conservation and pollution mitigation strategies to obtain the greatest biodiversity and ecological integrity. If we want to protect the integrity of all natural systems in an increasingly polluted world, interdisciplinary approaches drawing from toxicology, ecology, and environmental science are essential.

REFERENCES

1. Ali, H., & Khan, E. (2019). Trophic transfer, bioaccumulation, and biomagnification of non-essential hazardous heavy metals and metalloids in food chains/webs—Concepts and implications for wildlife and human health. *Human and Ecological Risk Assessment: An International Journal*, *25*(6), 1353–1376.
2. Alimba, C. G., & Faggio, C. (2019). Microplastics in the marine environment: Current trends in environmental pollution and mechanisms of toxicological profile. *Environmental Toxicology and Pharmacology*, *68*, 61–74.
3. Belden, J. (2020). Introduction to ecotoxicology. In An Introduction to Interdisciplinary Toxicology (pp. 381–393).
4. Bergek, S., Ma, Q., Vetemaa, M., Franzén, F., & Appelberg, M. (2012). From individuals to populations: Impacts of environmental pollution on natural eelpout populations. *Ecotoxicology and Environmental Safety*, *79*, 1–12.
5. Brooks, B. W. (2022). Precision Ecotoxicology and the Biodiversity Crisis. *Environmental Science & Technology Letters*, *9*(4), 245–246.
6. Chormare, R., & Kumar, M. A. (2022). Environmental health and risk assessment metrics with special mention to biotransfer, bioaccumulation and biomagnification of environmental pollutants. *Chemosphere*, *302*, 134836.
7. Clements, W. H., & Rohr, J. R. (2009). Community responses to contaminants: Using basic ecological principles to predict ecotoxicological effects. *Environmental Toxicology and Chemistry*, *28*(9), 1789–1800.
8. Cook, J. (1977). Environmental pollution by heavy metals. *International Journal of Environmental Studies*, *10*(4), 253–266.
9. Edwards, C. A. (n.d.). The Impact of Pesticides on the Environment. In The Pesticide Question (pp. 13–46). Boston, MA: Springer US.

10. Edwards, C. A. (2002). Assessing the effects of environmental pollutants on soil organisms, communities, processes and ecosystems. *European Journal of Soil Biology, 38*(3–4), 225–231.

11. Efroymson, R. A., & Murphy, D. L. (2001). Ecological risk assessment of multimedia hazardous air pollutants: estimating exposure and effects. *Science of The Total Environment, 274*(1–3), 219–230.

12. Fisken, A. G. (1972). Pesticides and the Environment. *Medicine, Science and the Law, 12*(2), 104–108.

13. Förstner, U., & Müller, G. (1973). Heavy metal accumulation in river sediments: A response to environmental pollution. *Geoforum, 4*(2), 53–61.

14. Goksøyr, A. (2006). Endocrine Disruptors in the Marine Environment: Mechanisms of Toxicity and their Influence on Reproductive Processes in Fish. *Journal of Toxicology and Environmental Health, Part A, 69*(1–2), 175–184.

15. Gola, D., Kumar Tyagi, P., Arya, A., Chauhan, N., Agarwal, M., Singh, S. K., & Gola, S. (2021). The impact of microplastics on marine environment: A review. *Environmental Nanotechnology, Monitoring & Management, 16*, 100552.

16. He, F., Liu, R., Tian, G., Qi, Y., & Wang, T. (2023). Ecotoxicological evaluation of oxidative stress-mediated neurotoxic effects, genetic toxicity, behavioral disorders, and the corresponding mechanisms induced by fluorene-contaminated soil targeted to earthworm (Eisenia fetida) brain. *Science of The Total Environment, 871*, 162014.

17. Hogg, J. C., & Van Eeden, S. (2009). Pulmonary and systemic response to atmospheric pollution. *Respirology, 14*(3), 336–346.

18. Karimi, B., Cahurel, J.-Y., Gontier, L., Charlier, L., Chovelon, M., Mahé, H., & Ranjard, L. (2020). A meta-analysis of the ecotoxicological impact of viticultural practices on soil biodiversity. *Environmental Chemistry Letters, 18*(6), 1947–1966.

19. Kazlauskiene, N., Svecevicius, G., Marciulioniene, D., Montvydiene, D., Kesminas, V., Staponkus, R., Taujanskis, E., & Sluckaite, A. (2012). The effect of persistent pollutants on aquatic ecosystem: A complex study. *2012 IEEE/OES Baltic International Symposium (BALTIC)*, 1–6.

20. Lavoie, R. A., Jardine, T. D., Chumchal, M. M., Kidd, K. A., & Campbell, L. M. (2013). Biomagnification of Mercury in Aquatic Food Webs: A Worldwide Meta-Analysis. *Environmental Science & Technology, 47*(23), 13385–13394.

21. Menezes-Oliveira, V. B., Scott-Fordsmand, J. J., Soares, A. M. V. M., & Amorim, M. J. B. (2014). Development of ecosystems to climate change and the interaction with pollution—Unpredictable changes in community structures. *Applied Soil Ecology, 75*, 24–32.

22. Mohammad Ali, M., Hossain, D., Al-Imran, Suzan Khan, M., Begum, M., & Hasan Osman, M. (2021). Environmental Pollution with Heavy Metals: A Public Health Concern. In Heavy Metals - Their Environmental Impacts and Mitigation. IntechOpen.

23. Muñoz, R., Malhautier, L., Fanlo, J.-L., & Quijano, G. (2015). Biological technologies for the treatment of atmospheric pollutants. *International Journal of Environmental Analytical Chemistry, 95*(10), 950–967.

24. Obiakor, M. O., Tighe, M., Pereg, L., & Wilson, S. C. (2017). Bioaccumulation, trophodynamics and ecotoxicity of antimony in environmental freshwater food webs. *Critical Reviews in Environmental Science and Technology, 47*(22), 2208–2258.

25. Pedro, S., Fisk, A. T., Tomy, G. T., Ferguson, S. H., Hussey, N. E., Kessel, S. T., & McKinney, M. A. (2017). Mercury and persistent organic pollutants in native and invading forage species of the Canadian Arctic: Consequences for food web dynamics. *Environmental Pollution, 229*, 229–240.

26. Pesce, S., Bérard, A., Coutellec, M.-A., Hedde, M., Langlais-Hesse, A., Larras, F., Leenhardt, S., Mongruel, R., Munaron, D., Sabater, S., & Gallai, N. (2023). Linking ecotoxicological effects on biodiversity and ecosystem functions to impairment of ecosystem services is a

challenge: an illustration with the case of plant protection products. *Environmental Science and Pollution Research, 32*(6), 2773–2785.

27. Rathi, B. S., Kumar, P. S., & Vo, D.-V. N. (2021). Critical review on hazardous pollutants in water environment: Occurrence, monitoring, fate, removal technologies and risk assessment. *Science of The Total Environment, 797*, 149134.

28. Rutgers, M. (2008). Field effects of pollutants at the community level — Experimental challenges and significance of community shifts for ecosystem functioning. *Science of The Total Environment, 406*(3), 469–478.

29. Samuel, P. O., Edo, G. I., Oloni, G. O., Ugbune, U., Ezekiel, G. O., Essaghah, A. E. A., & Agbo, J. J. (2023). Effects of chemical contaminants on the ecology and evolution of organisms a review. *Chemistry and Ecology, 39*(10), 1071–1107.

30. Shetty, S. S., D, D., S, H., Sonkusare, S., Naik, P. B., Kumari N, S., & Madhyastha, H. (2023). Environmental pollutants and their effects on human health. *Heliyon, 9*(9), e19496.

31. Streit, B. (1992). Bioaccumulation processes in ecosystems. *Experientia, 48*(10), 955–970.

32. Suzuki, T., Hidaka, T., Kumagai, Y., & Yamamoto, M. (2020). Environmental pollutants and the immune response. *Nature Immunology, 21*(12), 1486–1495.

33. Tang, Y., Yin, M., Yang, W., Li, H., Zhong, Y., Mo, L., Liang, Y., Ma, X., & Sun, X. (2019). Emerging pollutants in water environment: Occurrence, monitoring, fate, and risk assessment. *Water Environment Research, 91*(10), 984–991.

34. Valavanidis, A., Vlahogianni, T., Dassenakis, M., & Scoullos, M. (2006). Molecular biomarkers of oxidative stress in aquatic organisms in relation to toxic environmental pollutants. *Ecotoxicology and Environmental Safety, 64*(2), 178–189.

35. Vandana, Priyadarshanee, M., Mahto, U., & Das, S. (2022). Mechanism of toxicity and adverse health effects of environmental pollutants. In Microbial Biodegradation and Bioremediation (pp. 33–53). Elsevier.

36. Vandegehuchte, M. B., & Janssen, C. R. (2011). Epigenetics and its implications for ecotoxicology. *Ecotoxicology, 20*(3), 607–624.

37. Vasseur, P., & Cossu-Leguille, C. (2006). Linking molecular interactions to consequent effects of persistent organic pollutants (POPs) upon populations. *Chemosphere, 62*(7), 1033–1042.

38. Verhaar, H. J. M., van Leeuwen, C. J., & Hermens, J. L. M. (1992). Classifying environmental pollutants. *Chemosphere, 25*(4), 471–491.

39. Wasi, S., Tabrez, S., & Ahmad, M. (2013). Toxicological effects of major environmental pollutants: an overview. *Environmental Monitoring and Assessment, 185*(3), 2585–2593.

40. Wood, T. J., & Goulson, D. (2017). The environmental risks of neonicotinoid pesticides: a review of the evidence post 2013. *Environmental Science and Pollution Research, 24*(21), 17285–17325.

41. Xiang, Y., Jiang, L., Zhou, Y., Luo, Z., Zhi, D., Yang, J., & Lam, S. S. (2022). Microplastics and environmental pollutants: Key interaction and toxicology in aquatic and soil environments. *Journal of Hazardous Materials, 422*, 126843.

42. Zhang, B., Matchinski, E. J., Chen, B., Ye, X., Jing, L., & Lee, K. (2019). Marine Oil Spills—Oil Pollution, Sources and Effects. In World Seas: An Environmental Evaluation (pp. 391–406).

Chapter - 16

CONSERVATION OF BIODIVERSITY IN THE INDIAN KNOWLEDGE SYSTEM

Dr. Surendra Pal Singh [1*] and Dr. Pooja Kumari[2]

[*]Assistant Professor, Department of Teacher Education, D. S. College, Aligarh-202001, U.P., India

[2]Assistant Professor, Department of Zoology, D. S. College, Aligarh-202001, U.P., India

***Corresponding email: spsinghdscollege@gmail.com**

ABSTRACT

The Indian Knowledge System (IKS) offers a profound understanding of nature and biodiversity conservation, deeply rooted in traditional practices, religious beliefs, and cultural heritage. This study explores how ancient Indian texts, folklore, and indigenous practices reflect a harmonious relationship between humans and nature. Concepts like sacred groves, water conservation through traditional harvesting systems, and sustainable agricultural practices exemplify the ecological wisdom embedded in IKS. The role of Ayurveda and traditional healing systems further emphasizes the value of biodiversity. Additionally, the study highlights the contributions of local communities in preserving plant and animal species through indigenous knowledge passed down through generations. By integrating these traditional approaches with modern scientific methods, sustainable solutions can be developed to address contemporary environmental challenges. The research underscores the relevance of reviving and applying indigenous knowledge for biodiversity conservation, ensuring ecological balance, and promoting environmental stewardship. This holistic perspective serves as a valuable framework for policymakers, environmentalists, and researchers in the pursuit of sustainable development goals.

Keywords: *Biodiversity Conservation, Indian Knowledge System (IKS), Traditional Ecological Knowledge (TEK), Indigenous Practices, Ayurveda and Ethnobotany, Environmental Ethics*

1. Overview of the Conservation of Biodiversity

Biodiversity conservation (BC) involves protecting and managing the diversity of life on Earth, including species, ecosystems, and genetic resources, which are closely linked to human culture, spirituality, and well-being. Although traditional societies have long used sustainable practices to conserve biodiversity, rapid population growth and expanding economic activities have placed increasing pressure on natural ecosystems. Indigenous communities have historically protected large areas of biodiversity through traditional knowledge and management practices, which can serve as effective models for conservation (Mekonen, 2017). However, modern conservation strategies sometimes undermine these traditional systems. Therefore, integrating indigenous knowledge with modern approaches, such as co-management and bio-cultural conservation, can improve biodiversity protection. Biodiversity-rich regions like the Western Ghats demonstrate the importance of biodiversity for livelihoods and ecosystem services, highlighting the need to preserve both biological resources and indigenous knowledge systems (Negi et al., 2021).

2. Conventional Ecological Knowledge

Traditional Ecological Knowledge (TEK) plays an important role in biodiversity conservation by providing locally based, qualitative knowledge developed through long-term interaction with the environment. Unlike Western scientific knowledge, which often follows a reductionist and quantitative approach, TEK focuses on maintaining the health and balance of local ecosystems (Mekonen, 2017). In India, TEK is closely linked to cultural practices and community-based resource management that support biodiversity and help communities adapt to environmental and climate changes. Since this knowledge is shared collectively within communities, community-based conservation programs can effectively support biodiversity protection. Integrating indigenous knowledge with modern conservation strategies can lead to innovative solutions and help preserve both biological diversity and cultural heritage in the face of globalization and environmental challenges (Mekonen, 2017).

3. **Obstacles to the Preservation of Biodiversity**

Biodiversity refers to the variety and variability of living organisms within a particular habitat, including genetic, species, and ecosystem diversity. It plays a vital role in maintaining ecosystem stability, supporting ecosystem services, and ensuring human well-being. However, biodiversity is declining rapidly due to increasing human activities such as deforestation, habitat destruction, and unsustainable exploitation of natural resources. The loss of biodiversity disrupts ecological balance, weakens ecosystem services, and threatens the availability of valuable biological resources such as medicinal plants and other natural products used in healthcare and agriculture.

India is recognized as one of the world's seventeen mega-diverse countries and possesses enormous biological wealth. The country is home to more than 92,000 animal species and around 45,000 plant species. This rich biodiversity provides essential ecosystem services and supports the livelihoods of millions of people. Many plant species found in India possess important pharmacological properties and are widely used in traditional medicinal systems. Indigenous communities have historically maintained a close relationship with nature, developing knowledge systems that integrate biodiversity conservation with daily life and cultural traditions. Their agricultural practices, crop diversity, and environmental management strategies reflect a deep understanding of local ecological conditions.

Traditional approaches have helped maintain the balance between food production, climate conditions, and biodiversity. One example of such an approach is the concept of Agro Biodiversity Parks, where diverse crop varieties and medicinal plants are conserved and cultivated under natural ecological conditions. These systems help protect valuable plant species while promoting sustainable agriculture and environmental conservation. Such traditional practices demonstrate how communities have historically contributed to biodiversity conservation and resource management (Nulkar et al., 2021). However, when biodiversity degradation goes unrecognized, it can eventually lead to severe ecological damage and threaten both natural ecosystems and human survival.

i. **Loss of Habitat and Urbanization**

One of the most significant drivers of biodiversity loss is habitat destruction caused by rapid urbanization. As cities expand, natural habitats are often cleared for infrastructure development, industrial activities, and residential settlements. This leads to habitat fragmentation, which isolates species populations and disrupts ecological processes essential for maintaining ecosystem health. In India, rapid urban expansion has placed tremendous pressure on natural ecosystems and wildlife habitats.

Several Indian cities such as Thane, Lucknow, Kanpur, Chennai, Bangalore, Vijayawada, and Hyderabad have experienced significant habitat fragmentation due to uncontrolled urban growth. These cities are often located near biodiversity-rich regions, which makes the impact even more severe. For example, urban expansion near biodiversity hotspots such as the Western Ghats further accelerates ecosystem degradation. Habitat fragmentation not only threatens species survival but also affects ecological functions that are important for human health and environmental stability (Paul, 2016).

With projections suggesting that more than 500 million people will live in Indian cities by 2028, the challenge of balancing urban development with biodiversity conservation is becoming increasingly urgent. Sustainable urban planning, community participation, and environmentally responsible policies are necessary to ensure that biodiversity is protected while meeting the developmental needs of growing urban populations

ii. **Effects of Climate Change**

One of the most significant drivers of biodiversity loss is habitat destruction caused by rapid urbanization. As cities expand, natural habitats are often cleared for infrastructure development, industrial activities, and residential settlements. This leads to habitat fragmentation, which isolates species populations and disrupts ecological processes essential for maintaining ecosystem health. In India, rapid urban expansion has placed tremendous pressure on natural ecosystems and wildlife habitats.

Several Indian cities such as Thane, Lucknow, Kanpur, Chennai, Bangalore, Vijayawada, and Hyderabad have experienced significant habitat fragmentation due to uncontrolled urban growth. These cities are often located near biodiversity-rich regions, which makes the impact even more severe. For example, urban expansion near biodiversity hotspots such as the Western Ghats further accelerates ecosystem degradation. Habitat fragmentation not only threatens species survival but also affects ecological functions that are important for human health and environmental stability (Paul, 2016).

With projections suggesting that more than 500 million people will live in Indian cities by 2028, the challenge of balancing urban development with biodiversity conservation is becoming increasingly urgent. Sustainable urban planning, community participation, and environmentally responsible policies are necessary to ensure that biodiversity is protected while meeting the developmental needs of growing urban populations.

4. **Conservation's Philosophical Foundations**

The philosophical foundations of conservation emphasize the ethical principles that guide the protection of nature and biodiversity. These include recognizing the intrinsic value of nature, determining the extent of environmental protection, and understanding the relationship between humans and other living beings. Many cultural and philosophical traditions, particularly Eastern philosophies and indigenous belief systems, view nature as sacred and interconnected with human life. These perspectives promote environmental stewardship and respect for natural resources. Conservation ethics also address important issues such as sustainable resource use, social justice between rich and poor communities, and the link between biodiversity and human well-being. A holistic approach that integrates ethical values, cultural beliefs, and environmental responsibility is essential for achieving both biodiversity conservation and human welfare.

i. **Nature and Spirituality**

Spiritual beliefs have played an important role in biodiversity conservation throughout human history. Many cultures in Asia, Africa, and the Americas incorporate conservation ethics within their spiritual traditions. In India, sacred groves are a notable example of this relationship between spirituality and nature. These are small forest areas protected due to religious and cultural beliefs and serve as important habitats for diverse plant and animal species (Wickramasinghe, 2013). Sacred natural sites (SNS) represent areas where cultural traditions, spiritual values, and ecological conservation overlap. Since the 1992

Earth Summit, the concept of sustainable resource use has increasingly recognized the role of cultural diversity, traditional institutions, and indigenous environmental knowledge in biodiversity conservation. Biological and cultural diversity are closely interconnected, and protecting sacred sites also helps preserve indigenous traditions and values (K. Patel & K. Patel, 2013). Researchers highlight that spirituality can foster a deep connection with nature and encourage conservation behavior, offering valuable insights for modern biodiversity conservation strategies.

ii. **The Morality of Conservation**

The deliberate destruction of biodiversity raises serious ethical concerns because decisions affecting ecosystems and future generations are often made without their involvement (Spash & Aldred, 1998). Modern environmental philosophy has expanded traditional ethics to include responsibilities toward nature and non-human species, rather than focusing only on human interests. Three main ethical approaches guide biodiversity conservation. The first, animal ethics, emphasizes moral responsibilities toward sentient animals and recognizes their rights and capacity to feel pain. The second approach, ecological holism, argues that ecosystems, species, and the Earth itself possess intrinsic value and should be protected regardless of human interests. The third approach combines human-centered ethical theories, such as utilitarian and duty-based perspectives, to evaluate which species or ecosystems should be conserved. Together, these perspectives broaden ethical thinking and highlight humanity's moral responsibility to protect biodiversity and the natural environment.

5. India's Agricultural Biodiversity

Indian farmers have developed and maintained rich agrobiodiversity over thousands of years by selecting, conserving, and exchanging seeds, plant varieties, and livestock breeds. This diversity supports food security, livelihoods, and environmental sustainability for millions of rural families. However, India is experiencing a rapid decline in the genetic diversity of crops and livestock due to industrial agriculture, monocropping, hybrid varieties, chemical inputs, and global trade pressures. These changes threaten traditional farming systems, reduce soil fertility, cause deforestation, and increase water scarcity. Traditional agroecological farming practices, which maintain on-farm biodiversity and rely on indigenous knowledge, help improve ecosystem health and allow farmers to adapt to environmental changes (Trajkova et al., 2022). Protecting traditional crop varieties and integrating indigenous knowledge into agricultural policies is essential for sustainable agriculture, biodiversity conservation, and long-term food security in India.

6. **Conventional Farming Methods**

Traditional farming communities in India have developed agricultural practices based on deep knowledge of local ecosystems and climatic conditions. These practices, refined over generations, include crop rotation, intercropping, and organic farming, which help maintain soil fertility, ecological balance, and agricultural biodiversity (Negi et al., 2021). Tribal farmers also adapt their farming methods to changing weather patterns, demonstrating flexible and dynamic agroecological knowledge. For example, farmers in the mountainous regions of Kerala adjust the timing and spacing of tuber planting based on rainfall patterns to ensure stable production (T.R. et al., 2018). However, the spread of modern agricultural technologies, high-yielding varieties, and commercial farming has led to the decline of traditional crop diversity and indigenous knowledge systems. Economic reforms and large-scale agricultural projects often overlook the ecological benefits of traditional practices and may threaten sustainable farming systems. Efforts to promote sustainable agriculture should therefore recognize and support indigenous farming knowledge, which plays a crucial role in maintaining biodiversity, ecological stability, and resilient agricultural systems.

i. **Sustainability and Crop Diversity**

Biodiversity is essential for human survival and provides many important environmental services. In the Indian subcontinent, biodiversity is maintained through traditional cultural practices and sustainable resource management systems. Agricultural ecosystems play a major role in conserving biodiversity through crop production and the presence of wild crop relatives, animals, insects, birds, and other organisms that support ecological balance (Kavitha et al., 2016). Crop diversity has long been recognized as vital for food security and includes a wide range of plants such as cereals, pulses, fruits, vegetables, medicinal plants, spices, and fiber crops. Crop varieties are a key source of genetic diversity in agriculture. Different species and cultivars grow and mature at different times and show varying resistance to pests, diseases, and environmental conditions. This diversity helps reduce crop damage, improves productivity, and prevents environmental contamination. Therefore, maintaining a wide variety of crop species and cultivars is important for sustainable agriculture, ecological stability, and long-term food security.

7. Prospects for Biodiversity Conservation in the Future

Biodiversity conservation in India requires a multifaceted approach that integrates traditional knowledge with modern technologies and participatory institutions (Wickramasinghe, 2013). Tools like biotechnology and remote sensing can complement local ecological knowledge to monitor and protect ecosystems effectively. Combining these systems allows local communities to play an active role in management while benefiting from science-based strategies. Traditional knowledge offers long-term, flexible sustainability, whereas scientific institutions often work on fixed annual plans, creating an "out-of-phase" relationship. Building mutual trust between these knowledge systems is essential for successful conservation. When trust is established, collaborative strategies can be tested and scaled, ensuring biodiversity protection even in the face of urgent ecological crises.

8. **Novel Methods and Technologies**

Biodiversity conservation is increasingly embracing innovative technologies to improve efficiency, effectiveness, and inclusivity. Modern tools enable precise monitoring of ecosystems, real-time assessment of environmental health, and targeted interventions. Geographic Information Systems (GIS) are widely used to map ecosystems, track deforestation, and evaluate ecosystem conditions over time. Similarly, drone-based monitoring technologies are being deployed globally, offering rapid data collection on both abiotic and biotic ecosystem components. Complementary methods, including genetic, biochemical, and physical analyses, allow conservationists to study species, populations, and habitats in an integrated manner, enhancing the understanding of ecological dynamics.

Biotechnology is also playing a growing role in conservation. Applications include improving genetic diversity, developing resilient crop and animal varieties, restoring lost or endangered species, and rehabilitating degraded habitats. When applied carefully, biotechnological interventions can increase agricultural productivity, support climate adaptation strategies, and reduce pressures on forests and other natural ecosystems. At the same time, these technologies must be context-sensitive, adapted to local environmental, cultural, and socioeconomic conditions. For example, smallholder farmers may benefit from biofertilizers or other low-impact biotechnological interventions that complement traditional ecological methods for soil restoration and landscape management. Integrating modern biotechnologies with indigenous knowledge systems can create a balanced approach that leverages both scientific innovation and centuries of ecological understanding (Lahoz-Monfort & Magrath, 2021).

However, unregulated or poorly monitored use of technology carries significant risks. Mismanagement of biotechnological tools can cause ecological harm, disrupt local socio-economic systems, and exacerbate inequalities. Many local communities lack the infrastructure, training, and financial resources to effectively implement and govern advanced technological interventions. Without proper support, these communities risk dependency on external agencies, reducing their control over natural resources and undermining local conservation practices. Additionally, locally relevant tools are often limited by factors such as the absence of renewable energy sources, inadequate government support, and insufficient investment, slowing the adoption of new technologies.Socio-cultural and ethical considerations further complicate technological interventions in biodiversity conservation. Automated systems and high-tech solutions can be inflexible, failing to account for local knowledge, cultural frameworks, and social values. Conservation technologies that do not align with community practices may alienate local populations, preventing them from benefiting fully from the data and insights produced. Effective conservation strategies require technology to support rather than replace indigenous knowledge and local ecological expertise. Clear, intuitive, and accessible technological tools are essential, ensuring that communities remain active participants in conservation efforts while maintaining governance over their natural resources.

9. **Joint Conservation Initiatives**

Collaborative conservation approaches involve partnerships among local communities, governments, NGOs, the private sector, and other stakeholders to manage natural resources more effectively (Knowles Notin, 2006). Unlike exclusionary conservation, which often displaced local populations and failed to prevent illegal activities, collaborative models engage communities in planning, decision-making, and enforcement. Local knowledge, such as traditional rules restricting fishing or access to sacred forests, helps regulate resource use, reduce management costs, and enhance biodiversity protection. In many cases, community-managed areas have proven more effective than state-controlled reserves, demonstrating the value of inclusive, participatory conservation strategies.

10. Conclusion

Becauseitisrootedineco-religiousbeliefs,rituals,andpractices—such as tree worship, purification rites, earth worship, songs about forests, the use of medicinal plants, traditional ecological practices, and local resource management—the traditional ecological knowledge (TEK) system has proven effective. Hindu environmental ethics, which strongly emphasize the protection of forests, mountains, trees, and land, encompass these practices. Ancient Hindu kingdoms in the Indian subcontinent issued directives, in the form of injunctions, to address challenges posed by urbanization, industrialization, overpopulation, and climate change. Examples of such directives include Tabedari, Kshekawasikarani, self-imposed bans, and Dharmaprastha, which restricted hunting, tree-felling, and other environmentally harmful activities.There are even historical accounts of curses imposed on those who illegally cut trees or misused rivers and forests while expanding territories. Although these directives originally arose from practical survival needs, they have the potential to reinforce conservation practices among specific populations, such as forest-dwellers and hill ethnic groups. To develop and implement innovative, economical and practical forestry conservation strategies on a wider scale, it is essential to follow these practices rigorously using proven methodologies, rather than merely defending the customs or focusing solely on their religious and cultural meanings concerning the conservation of trees, forests, rivers, or mountains. In Nepal, where biodiversity is declining at an alarming rate, such an approach could provide a viable means to revive the forestry conservation practices of certain hill ethnic and forest communities (Mekonen, 2017).

REFERENCES

1. Mekonen, S. (2017). Traditional ecological knowledge's contributions to biodiversity preservation.
2. S. D. Bhatt, I., Thakur, S., K. Joshi, R., Pathak, R., Negi, V., & S. Rawal, R. (2021). Assessing the Need for Indigenous Knowledge to Be Mainstreamed for Sustainable Bioresource Use in the Indian Himalayan Region. nih.gov/ncbi.nlm
3. Ghate, K., Nulkar, S., Bedarkar, M., and Nulkar, G. (2021). Microbes that hitchhike: a decline in biodiversity and the emergence of zoonoses. nih.gov/ncbi.nlm
4. Paul, S. (2016). Urbanization's Effect on Green Spaces: An Examination of the Human Factors Influencing Delhi's Land Use and Land Cover Change.
5. V. Kumar and K. Chopra, A. (2009). An overview of how climate change is affecting India's biodiversity, particularly in the Himalayan area.
6. Hagerman, M. S. & Chan, K. MA. (2009). Impacts, adaption techniques, and future research directions related to biodiversity conservation and climate change. nih.gov/ncbi.nlm
7. K. Patel, M. and P. Patel. (2013). In Gujarat, India's Banaskantha area, sacred groves contribute to the preservation of plant variety. [PDF]
8. L. Splash, C. & Aldred, J. (1998). conservation of wildlife.
9. Arsov, S., Koleva Gudeva, L., and Trajkova, F. (2022). The function and significance of biodiversity in agriculture.
10. Joseph, J., N., T.R., S., E., R., K.K., S., and P. N. Anilkumar. (2018). Decentralized conservation of agrobiodiversity: A multi-stakeholder participatory study.

11. Soumitra, P., Kavitha, K., and Padmaja, R. (2016). Recognizing how household dietary diversity and crop diversity are related in India's semi-arid regions.

12. Lahoz-Monfort, J. & Magrath, M. J. L. (2021). An Extensive Review of Technologies for Monitoring and Conserving Species and Habitat. nih.gov/ncbi.nlm

13. K. Knowles Notin. (2006). The Kaw Nature Reserve in French Guiana serves as a case study for the concepts of learning to change and changing learning in environmental management.

Chapter - 17

FROM FIBROSIS TO CIRRHOSIS: MECHANISMS AND CLINICAL IMPLICATIONS

Zubaida Khatoon[1], Harikesh Kumar Singh[1], Alok Yadav[1], Dr. Hadiya Husain[1*] and Dr. Shalini Singh[2]

[1]Department of Zoology, Faculty of Science, University of Lucknow, Lucknow-226007

[2]Government Degree College Kuchlai, Sitapur-261303

Corresponding email: hadiya.husain@gmail.com

ABSTRACT

From Fibrosis to Cirrhosis: Mechanisms and clinical implications explore every stage of liver disease from early fibrosis to advanced cirrhosis. This chapter discusses key pathophysiological processes such as hepatic damage, inflammatory signaling and hepatic stellate cell (HSC) activation, as well as major molecular pathways and etiologic variables. It examines histological and clinical changes as the disease progresses, with an emphasis on decompensating and hemodynamic alterations. Nanoparticles play an important role in diagnosis, staging and treatment, as do both invasive and non-invasive diagnostic technologies. The clinical implications, including prognosis, hepatocellular carcinoma risk, and liver transplantation, are discussed. The chapter finishes with current therapeutic techniques and future research goals that focus on nanotechnology and precision medicine.

Keywords: Liver fibrosis, Cirrhosis, Hepatic stellate cells, Inflammatory signaling, Nanoparticles, Precision medicine.

1. INTRODUCTION

Chronic liver diseases (CLDs) represent a major global health burden, responsible for millions of deaths each year and significantly contributing to worldwide illness and disability (Asrani et al., 2019). These diseases can develop due to various causes, such as viral hepatitis, chronic alcohol use, non-alcoholic fatty liver disease (NAFLD) and autoimmune disorders. All these conditions can cause continuous damage to the liver. A central feature of long-term liver injury is fibrosis, which is marked by the excessive build-up and altered organization of extracellular matrix (ECM) components-particularly collagen types I and III (Bataller & Brenner, 2005). This build-up disrupts the liver's normal structure and reduces its ability to perform vital functions.

Importantly, liver fibrosis is a dynamic condition and can be reversed in its early stages if the underlying cause is removed or treated effectively (Kisseleva & Brenner, 2021). However, ongoing liver injury may lead to cirrhosis, a severe stage of liver damage defined by extensive scarring, the formation of regenerative nodules, and disordered blood flow through the liver (Friedman, 2008). This stage is associated with serious complications, including portal hypertension, liver failure and a higher risk of liver cancer, such as hepatocellular carcinoma (HCC) (Tsochatzis et al., 2014). At this point, liver damage is often considered irreversible, and clinical focus shifts toward symptom management and potential liver transplantation.

Emerging advancements in nanotechnology have introduced new opportunities for improving liver disease treatment. Nanoparticles (NPs), which range in size from 1 to 100 nanometers, offer unique physical and chemical properties that make them well-suited for delivering drugs to specific liver tissues, enhancing imaging techniques, and modulating disease processes at a molecular level (Peer et al., 2020). In cases of liver fibrosis and cirrhosis, NPs can be tailored to accumulate in fibrotic areas, which increase treatment accuracy while reducing side effects throughout the body (Danhier et al., 2012). These developments are not only helping scientists better understand how liver fibrosis develops but also opening new paths to potentially stop or even reverse its progression.

FROM FIBROSIS TO CIRRHOSIS

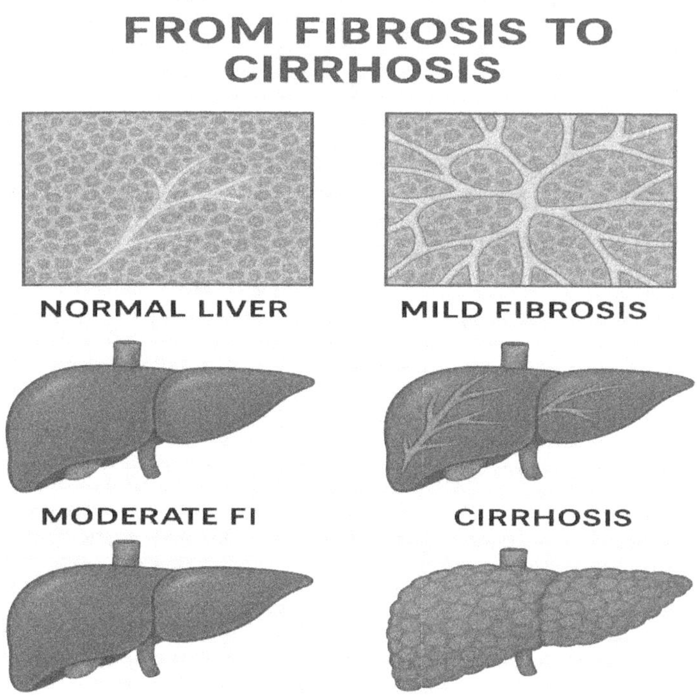

Figure1. Stages of Liver Fibrosis

2. PATHOPHYSIOLOGY OF LIVER FIBROSIS

2.1. Hepatic Injury and Inflammatory Signaling

Chronic liver damage can result from several underlying factors such as viral infections, long-term alcohol use, metabolic imbalances, and autoimmune diseases. These conditions trigger a persistent inflammatory response, which plays a central role in the development of fibrosis (Bataller & Brenner, 2005). A key player in this process is the liver's own immune cells, known as Kupffer cells. These cells detect injury and release inflammatory molecules like TNF-α, IL-1β, and IL-6, which in turn attract and activate other immune cells. This leads to a chain reaction that ultimately activates hepatic stellate cells (HSCs), which are primarily responsible for fibrosis (Jager et al., 2016). Prolonged inflammation fosters an environment that favors scar tissue formation and sustains liver injury.

Recent advances in nanotechnology have made it possible to design nanoparticles (NPs) that specifically target Kupffer cells to control inflammation and encourage tissue repair. By manipulating their size, charge and surface properties, these NPs can be customized to effectively interact with liver macrophages. The local environment within the liver also influences how these NPs are absorbed and how they function. Although concerns remain about their toxicity and how the body clears them, early research shows promise for NPs in treating liver conditions, including fibrosis, infections and cancer (Colino et al., 2020).

2.2. Activation of Hepatic Stellate Cells (HSCs)

Hepatic stellate cells are found in the space of Disse and normally function in storage of vitamin A. However, when the liver is injured or inflamed, these cells become activated and transform into fibrogenic cells that behave like myofibroblasts. Once activated, they multiply and produce large quantities of extracellular matrix (ECM) proteins, particularly collagen types I and III, which are central to the formation of liver scar tissue (Friedman, 2008). This process is regulated by various chemical signals, with transforming growth factor-beta (TGF-β) being the primary molecule that stimulates stellate cell activation and collagen production through Smad-related and other pathways (Dooley & ten Dijke, 2012).

2.3. Key Molecular Pathways

Several signaling pathways inside liver cells regulate the behavior and activation of HSCs during fibrosis. Chief among these is the TGF-β/Smad pathway, which directly promotes the production of fibrotic proteins

(Friedman, 2008). Additionally, the Wnt/β-catenin pathway contributes to HSC growth and survival, while the Hedgehog signaling pathway influences both fibrosis and the liver's ability to regenerate (Machado et al., 2018).

3. ETIOLOGIES AND TRIGGERS OF FIBROSIS

Liver fibrosis is the shared outcome of many chronic liver diseases, each with different underlying causes and disease mechanisms. Major contributors include viral hepatitis, alcoholic liver disease (ALD), NAFLD, autoimmune and cholestatic liver diseases, as well as hereditary and metabolic disorders. Although antiviral treatments have improved, viral hepatitis still causes a significant health burden, especially in areas with limited healthcare infrastructure.

In ALD, chronic alcohol use damages liver cells and disrupts normal metabolic functions. Toxic alcohol byproducts, such as acetaldehyde, generate oxidative stress and promote inflammation, both of which accelerate fibrosis (Seitz & Stickel, 2010). In this setting, Kupffer cells and HSCs are activated, leading to scar tissue accumulation. Fibrosis in NAFLD arises from the buildup of fat in the liver, increased oxidative stress, and inflammation, which together damage liver cells and activate fibrotic pathways.

Inherited metabolic diseases such as hemochromatosis, Wilson's disease, and alpha-1 antitrypsin deficiency contribute to fibrosis by causing toxic build-up of metals or faulty proteins. These triggers create ongoing oxidative stress and liver cell damage, driving the progression toward cirrhosis (Malhi & Gores, 2008). Interestingly, nanotechnology-based drug delivery systems can enhance the effectiveness of antioxidants and anti-inflammatory treatments used in ALD by improving their bioavailability. This strategy helps minimize liver injury and fibrotic progression, highlighting the growing role of nanomedicine in tailored therapies for liver disease.

4. PROGRESSION TO CIRRHOSIS

4.1. Histological and Clinical Definitions

Cirrhosis is the end-stage of chronic liver disease and is marked by extensive fibrosis and the formation of regenerative nodules that disrupt the liver's normal architecture (Schuppan & Afdhal, 2008). Under the microscope, cirrhosis shows thick bands of scar tissue enclosing nodules of liver cells, which distorts the structure of liver lobules and blood vessels. Clinically, patients with cirrhosis may initially show few symptoms (compensated stage) but eventually progress to severe liver dysfunction (decompensated stage) with potentially life-threatening complications (Tsochatzis et al., 2014).

4.2. Structural and Hemodynamic Changes

As fibrosis develops into cirrhosis, the liver's structure changes significantly, particularly its blood vessels. The growing scar tissue and regenerative nodules compress liver vessels, increasing resistance to blood flow and resulting in portal hypertension-a key sign of cirrhosis where blood pressure rises in the portal vein system (Garcia-Tsao et al., 2017).

4.3. Decompensated Cirrhosis

In decompensated cirrhosis, the liver's functional reserve is overwhelmed, leading to complications such as jaundice, fluid buildup (ascites), confusion (hepatic encephalopathy), bleeding from varices, and kidney dysfunction (hepatorenal syndrome). These events are often set off by infections, internal bleeding, continued alcohol use, or drug-related liver injury. Treatment includes managing symptoms, identifying and treating triggers, and preventing further complications. Early diagnosis and a comprehensive care strategy are essential to improving survival chances (Mansour et al., 2018). Without timely intervention, patients with decompensated cirrhosis face a high risk of death unless they receive a liver transplant.

4.4. Nanoparticles in Cirrhosis Diagnosis

Early detection of cirrhosis is crucial for effective treatment. However, standard imaging techniques such as ultrasound and MRI may miss subtle early-stage changes. Nanoparticle-based contrast agents are emerging as valuable tools to improve imaging accuracy. When modified with magnetic or acoustic elements, these nanoparticles enhance the clarity of MRI and ultrasound images, making it easier to detect early fibrotic changes and measure liver stiffness (Zhang et al., 2023). These advanced imaging agents not only target fibrotic tissue more precisely but also offer real-time visualization, improving the ability to diagnose and track the progression of cirrhosis.

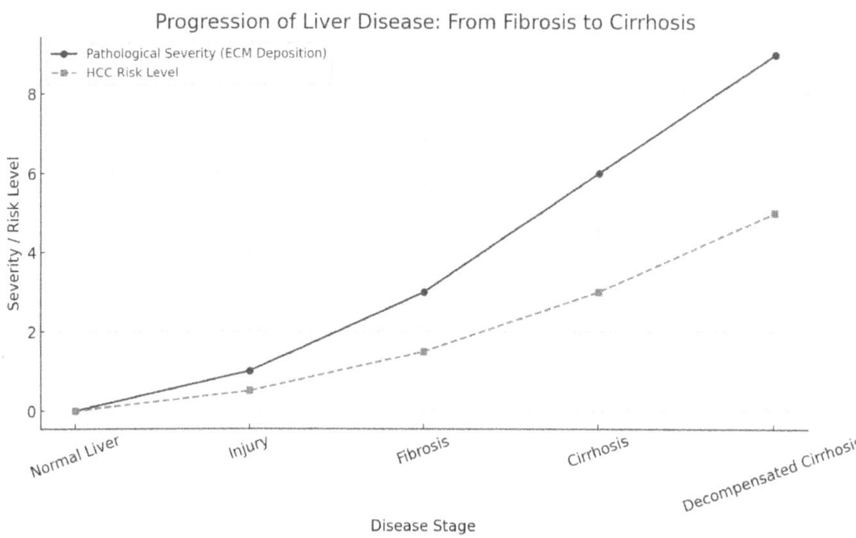

Figure 3. Schematic representation of liver disease progression from normal histology to decompensated cirrhosis. The curve reflects increasing extracellular matrix (ECM) deposition and pathological severity. A parallel increase in hepatocellular carcinoma (HCC) risk is shown in red (Friedman, 2008; Bataller & Brenner, 2005)

5. DIAGNOSTIC AND STAGING TOOLS

5.1. Histological Grading and Staging

The most reliable way to assess the severity and stage of liver fibrosis or cirrhosis is through examining liver tissue under a microscope. Two widely recognized scoring systems for this purpose are the METAVIR and Ishak systems. METAVIR categorizes fibrosis from F0 (no fibrosis) to F4 (cirrhosis), while the Ishak scale offers a more detailed evaluation ranging from 0 to 6 (Bedossa & Poynard, 1996; Ishak et al., 1995). Despite their accuracy, liver biopsies come with challenges such as being invasive, subject to sampling error, and influenced by differences in interpretation between observers. These limitations have encouraged the development of non-invasive alternatives.

5.2. Non-Invasive Methods

Non-invasive techniques have become increasingly popular because they are safer and more convenient. One such method is transient elastography (FibroScan®), which measures liver stiffness to estimate fibrosis severity and can be performed quickly at the bedside. It has been validated across various types of liver disease and correlates well with biopsy results (Castera et al., 2019).

Blood-based tests also provide useful insights. For example, the AST to Platelet Ratio Index (APRI) and the Fibrosis-4 (FIB-4) index use common laboratory values to estimate fibrosis. These tools are affordable and accessible, though they may be less precise in identifying moderate fibrosis (Vallet-Pichard et al., 2007).

5.3. Nanoparticle-Enhanced Imaging and Biomarkers

Nanotechnology is offering groundbreaking improvements in liver disease diagnostics. Nanoparticles linked to specific biological targets—such as activated hepatic stellate cells or components of the extracellular matrix—can significantly enhance imaging techniques like MRI, CT, and ultrasound by improving contrast and accuracy (Chen et al., 2022). For instance, iron oxide nanoparticles attached to collagen-binding molecules have shown the ability to detect early fibrotic changes via MRI. Integrating these technologies could enable earlier diagnosis, more accurate disease staging, and better planning of treatment strategies.

6. CLINICAL IMPLICATIONS OF FIBROSIS AND CIRRHOSIS

6.1. Prognosis and Outcomes

The extent of liver fibrosis is a major factor in predicting patient outcomes. More advanced fibrosis and cirrhosis are closely associated with serious complications such as portal hypertension, liver failure, and increased infection risk (Friedman, 2003). Long-term studies have shown that patients with higher levels of fibrosis have poorer survival rates, which underscores the importance of early diagnosis and timely treatment (D'Amico et al., 2006).

6.2. Risk of Hepatocellular Carcinoma

Cirrhosis is the most significant risk factor for developing hepatocellular carcinoma (HCC), accounting for up to 90% of all HCC cases worldwide (El-Serag, 2012). The fibrotic and inflamed environment in a cirrhotic liver fosters genetic damage and the formation of new blood vessels, processes that support cancer development. Regular monitoring is therefore essential to catch HCC at an early, more treatable stage (Friedman et al., 2004).

6.3. Liver Transplant Considerations

For patients with severe liver disease or decompensated cirrhosis, liver transplantation remains the only curative option. The Model for End-Stage Liver Disease (MELD) score is used to rank patients based on disease severity and risk of death (Kamath et al., 2001). While transplantation improves survival, it is limited by organ shortages and the possibility of post-transplant complications. Recurring fibrosis is a major cause of long-term graft failure, often driven by the return of the original disease, new liver conditions (such as autoimmune hepatitis or NAFLD), or chronic rejection (Berumen et al., 2020). In hepatitis C virus (HCV) patients, fibrosis can return and progress even faster after transplant (Berenguer et al., 2000).

6.4. Nanoparticles in Therapy and Monitoring

Multifunctional nanoparticles offer the dual benefit of delivering medication precisely to the liver while also enabling real-time monitoring of treatment responses through imaging. This approach improves treatment accuracy, minimizes systemic side effects, and allows for dynamic tracking of disease progression or regression.

7. THERAPEUTIC APPROACHES

7.1. Etiology-Specific Treatments

Managing liver fibrosis starts with treating the root cause. For example, antiviral drugs have proven effective in controlling hepatitis B and C, which helps halt the processes that lead to fibrosis (Marcellin et al., 2016).

7.2. Nanoparticle-Based Antifibrotic Strategies

Nanotechnology enhances the delivery of antifibrotic drugs by concentrating them directly in the liver, which boosts their effectiveness and reduces side effects elsewhere in the body. Delivery systems such as liposomes, polymer-based nanoparticles, and metal-based nanomaterials are being developed to achieve liver-specific drug targeting (Kim et al., 2022).

8. REVERSIBILITY AND REGRESSION OF FIBROSIS

8.1. Evidence for Reversibility

Contrary to earlier assumptions, liver fibrosis is not always permanent. When the underlying cause is removed, especially in early stages, the liver can heal. This process involves breaking down the excess extracellular matrix, clearing activated stellate cells, and restoring normal liver architecture (Kisseleva & Brenner, 2021; Iredale, 2007). Ongoing research is focused on identifying which factors influence this healing process and finding new drug targets to enhance it (Friedman & Bansal, 2006).

8.2. Nanoparticles in Promoting Fibrosis Regression

Among various nanoparticle systems, liposomes, inorganic particles, and nanomicelles are the most studied for delivering antifibrotic therapies. Liposome-based formulations currently in clinical trials represent a significant step forward in the treatment of liver fibrosis (Surendran et al., 2017). These particles improve how drugs are absorbed and directed to the liver, increasing their effectiveness while reducing side effects. This approach not only helps stop the progression of fibrosis but may also reverse it in some cases.

9. FUTURE PERSPECTIVES AND RESEARCH DIRECTIONS

Looking ahead, personalized nanomedicine is expected to play a larger role in liver disease management. By analyzing a patient's genetic, protein, and metabolic profile, researchers can design customized nanoparticles that specifically target the fibrotic pathways active in that individual (Bai X et al., 2020). This precision approach could significantly improve treatment outcomes while minimizing risks. Collaboration among scientists, clinicians, and regulatory agencies will be key to bringing these innovations from the lab into routine clinical use.

10. CONCLUSION

Liver fibrosis and cirrhosis are complex conditions with serious health consequences. Detecting and treating them early is essential to prevent permanent liver damage and related complications. Nanotechnology is rapidly transforming the way we diagnose and manage these diseases—offering tools for targeted therapy, enhanced imaging, and real-time monitoring.

The ability of nanomedicine to fine-tune treatment delivery, reduce side effects, and support personalized care marks a significant leap forward. Continued research and clinical application of nanoparticle-based solutions are crucial to fully realize their potential in improving patient outcomes in chronic liver disease.

REFERENCES

1. Asrani, S. K., Devarbhavi, H., Eaton, J., & Kamath, P. S. (2019). Burden of liver diseases in the world. *Journal of hepatology, 70*(1), 151-171.
2. Bai X, Su G, Zhai S. Recent Advances in Nanomedicine for the Diagnosis and Therapy of Liver Fibrosis. Nanomaterials (Basel). 2020 Sep 29;10(10):1945.
3. Bataller, R., & Brenner, D. A. (2005). Liver fibrosis. *The Journal of clinical investigation, 115*(2), 209-218.
4. Bedossa, P., & Poynard, T. (1996). An algorithm for the grading of activity in chronic hepatitis C. *Hepatology, 24*(2), 289-293.
5. Berenguer, M., Prieto, M., Rayón, J. M., Mora, J., Pastor, M., Ortiz, V.,... & Berenguer, J. (2000). Natural history of clinically compensated hepatitis C virus–related graft cirrhosis after liver transplantation. *Hepatology, 32*(4), 852-858.
6. Berumen, J., Baglieri, J., Kisseleva, T., & Mekeel, K. (2021). Liver fibrosis: Pathophysiology and clinical implications. *WIREs mechanisms of disease, 13*(1), e1499.
7. Castera, L., Friedrich-Rust, M., & Loomba, R. (2019). Noninvasive assessment of liver disease in patients with nonalcoholic fatty liver disease. *Gastroenterology, 156*(5), 1264-1281.
8. Colino, C. I., Lanao, J. M., & Gutierrez-Millan, C. (2020). Targeting of hepatic macrophages by therapeutic nanoparticles. *Frontiers in immunology, 11*, 218.
9. D'Amico, G., Garcia-Tsao, G., & Pagliaro, L. (2006). Natural history and prognostic indicators of survival in cirrhosis: a systematic review of 118 studies. *Journal of hepatology, 44*(1), 217-231.
10. Danhier, F., Ansorena, E., Silva, J. M., Coco, R., Le Breton, A., & Préat, V. (2012). PLGA-based nanoparticles: an overview of biomedical applications. *Journal of controlled release, 161*(2), 505-522.

11. Dooley, S., & Ten Dijke, P. (2012). TGF-β in progression of liver disease. *Cell and tissue research*, *347*(1), 245-256.

12. El-Serag, H. B. (2012). Epidemiology of viral hepatitis and hepatocellular carcinoma. *Gastroenterology*, *142*(6), 1264-1273.

13. Friedman, S. L. (2003). Liver fibrosis–from bench to bedside. *Journal of hepatology*, *38*, 38-53.

14. Friedman, S. L. (2004). Mechanisms of disease: mechanisms of hepatic fibrosis and therapeutic implications. *Nature clinical practice Gastroenterology & hepatology*, *1*(2), 98-105.

15. Friedman, S. L., & Bansal, M. B. (2006). Reversal of hepatic fibrosis—fact or fantasy?. *Hepatology*, *43*(S1), S82-S88.

16. Garcia-Tsao, G., Sanyal, A. J., Grace, N. D., Carey, W., & Practice Guidelines Committee of the American Association for the Study of Liver Diseases, the Practice Parameters Committee of the American College of Gastroenterology. (2007). Prevention and management of gastroesophageal varices and variceal hemorrhage in cirrhosis. *Hepatology*, *46*(3), 922-938.

17. Ishak, K., Baptista, A., Bianchi, L., Callea, F., De Groote, J., Gudat, F.,... & Thaler, H. (1995). Histological grading and staging of chronic hepatitis. *Journal of hepatology*, *22*(6), 696-699.

18. Jager, J., Aparicio-Vergara, M., & Aouadi, M. (2016). Liver innate immune cells and insulin resistance: the multiple facets of Kupffer cells. *Journal of internal medicine*, *280*(2), 209-220.

19. Kamath, P. S., Wiesner, R. H., Malinchoc, M., Kremers, W., Therneau, T. M., Kosberg, C. L.,... & Kim, R. W. (2001). A model to predict survival in patients with end–stage liver disease. *Hepatology*, *33*(2), 464-470.

20. Kim, J., Yun, Y. H., & Jung, Y. (2022). Therapeutic Potential of MicroRNAs and Their Nanoparticle-based Delivery in the Treatment of Liver Fibrosis. In *Biotechnologies for Gene Therapy: RNA, CRISPR, Nanobots, and Preclinical Applications* (pp. 1-23). Cham: Springer International Publishing.

21. Li, S., Li, H., Xu, X., Saw, P. E., & Zhang, L. (2020). Nanocarrier-mediated antioxidant delivery for liver diseases. *Theranostics, 10*(3), 1262.

22. Machado, M. V., & Diehl, A. M. (2018). Hedgehog signalling in liver pathophysiology. *Journal of hepatology, 68*(3), 550-562.

23. Malhi, H., & Gores, G. J. (2008). Cellular and molecular mechanisms of liver injury. *Gastroenterology, 134*(6), 1641-1654.

24. Mansour, D., & McPherson, S. (2018). Management of decompensated cirrhosis. *Clinical Medicine, 18*(2), s60-s65.

25. Marcellin, P., Gane, E., Buti, M., Afdhal, N., Sievert, W., Jacobson, I. M.,... & Heathcote, E. J. (2013). Regression of cirrhosis during treatment with tenofovir disoproxil fumarate for chronic hepatitis B: a 5-year open-label follow-up study. *The Lancet, 381*(9865), 468-475.

26. Peer, D., Karp, J. M., Hong, S., Farokhzad, O. C., Margalit, R., & Langer, R. (2020). Nanocarriers as an emerging platform for cancer therapy. *Nano-enabled medical applications*, 61-91.

27. Poilil Surendran, S., George Thomas, R., Moon, M. J., & Jeong, Y. Y. (2017). Nanoparticles for the treatment of liver fibrosis. *International journal of nanomedicine*, 6997-7006.

28. Schuppan, D., & Afdhal, N. H. (2008). Liver cirrhosis. *The Lancet, 371*(9615), 838-851.

29. Seitz, H. K., & Stickel, F. (2007). Molecular mechanisms of alcohol-mediated carcinogenesis. *Nature Reviews Cancer, 7*(8), 599-612.

30. Tsochatzis, E. A., Bosch, J., & Burroughs, A. K. (2014). Liver cirrhosis. *The Lancet, 383*(9930), 1749-1761.

31. Vallet-Pichard, A., Mallet, V., Nalpas, B., Verkarre, V., Nalpas, A., Dhalluin-Venier, V.,... & Pol, S. (2007). FIB-4: an inexpensive and accurate marker of fibrosis in HCV infection. comparison with liver biopsy and fibrotest. *Hepatology, 46*(1), 32-36.

32. Wu, Y., Zhang, J., He, W., Li, C., & Wang, Y. (2023). Nanomaterials for Targeting Liver Disease: Research Progress and Future Perspectives. *Nano Biomedicine & Engineering, 15*(2).

33. Zhang, J., Ning, Y., Zhu, H., Rotile, N. J., Wei, H., Diyabalanage, H.,... & Bawendi, M. G. (2023). Fast detection of liver fibrosis with collagen-binding single-nanometer iron oxide nanoparticles via T 1-weighted MRI. *Proceedings of the National Academy of Sciences, 120*(18), e2220036120.

www.ingramcontent.com/pod-product-compliance
Lightning Source LLC
LaVergne TN
LVHW052037200926
848481LV00005B/1712